MECHANICAL SEALS

Ready Engineering | Re-Engineering | Reverse Engineering

Dr. Ashok Punde

MECHANICAL SEALS

Ready Engineering | Re-Engineering | Reverse Engineering

Industrial Practices and Engineering Intelligence in Mechanical Seal Technology

---

Dr. Ashok Punde

Doctorate in Design Thinking & Innovation Management, in Mechanical Seals & Seal Support Systems.

48 Years in Mechanical Sealing Technology

Former MD / CEO — Mechanical Sealing Industry [25 Years]

Independent Consultant — Static & Dynamic Sealing Technology

This page has been left blank intentionally.

Primary Philosophy Statement

Ready Engineering prepares.
Re-Engineering improves.
Reverse Engineering discovers.
Without Engineering Intelligence,
Engineering becomes repetition.
With Engineering Intelligence,
Engineering becomes innovation.

Dr. Ashok Punde

This page has been left blank intentionally.

Dedication

To My Beloved Wife, Suman,

Your quiet strength has been the *steady axis* of my life,
holding me firm through both calm waters and uncertain tides.
As we began our journey together, I also began my life's work with Mechanical Seals and in many ways, both journeys have been shaped, sustained, and strengthened by you.

This book, born of Engineering, Science, Technology, and touched by Spirituality and Symbolism, is equally born of your unwavering support, your patience, and your silent sacrifices.

In every page lives a reflection of your presence.
In every thought, an echo of your encouragement.
In every achievement, a quiet testimony to your enduring faith in me.

You have been my balance in complexity, my calm in turbulence, and my strength in moments of doubt.
More than a companion, you have been the foundation upon which this journey stands, graceful, resilient, & ever-Giving.
With profound gratitude, love, and reverence, I dedicate this work to you.

With My Heartfelt Wishes & Blessings

Ashok

Engineering Knowledge Ladder

The Progressive Flow of Engineering Capability

Foreword

To All Readers,

Warm Greetings,

In today's evolving industrial landscape—where reliability, efficiency, and precision are paramount—mechanical sealing remains both a science and an art, shaped as much by real-world experience as by engineering theory. It is the essential balance that this book seeks to present with clarity and purpose.

Mechanical Seals: Ready Engineering, Re-Engineering, Reverse Engineering is not merely a technical text, but the distillation of nearly five decades of hands-on engineering experience. The insights shared herein are drawn from plant-floor realities, refined through leadership, and validated through sustained practical application.

Unlike conventional literature that often remains confined to ideal conditions, this work addresses the true complexities of industrial operations. It speaks directly to engineers and professionals who understand that factors such as installation practices, alignment, and operating conditions frequently influence performance more than design intent alone.

A central theme of the book is the redefinition of reverse engineering—not as mere duplication, but as an intelligent, disciplined process grounded in thermal, hydrodynamic, and material understanding. This perspective aims to elevate industry practice from replication to meaningful innovation.

The structured progression of Ready Engineering, Re-Engineering, and Reverse Engineering provides a practical framework—an engineering knowledge ladder—that supports both learning and mentorship. It is designed to help young engineers build intuition while enabling experienced professionals to transfer knowledge effectively.

Equally important is the book's candid acknowledgment that most seal failures originate not from design flaws, but from installation, misalignment, and operational realities—an insight that encourages a more responsible and holistic engineering approach.

In addressing Artificial Intelligence, the book maintains a balanced view, positioning it as a powerful enabler while reaffirming the irreplaceable role of human judgment, physical insight, and professional accountability.

It is my sincere hope that this work serves as a practical guide, a reference, and a source of thoughtful reflection for engineers across all levels.

A Humble Note from the Author

This being my first book, I sincerely seek the understanding and goodwill of readers for any unintentional errors or shortcomings that may have remained despite my best efforts.

Although I have previously written more than 1500 pages in article form, transforming experience and knowledge into a book has itself been a valuable learning journey. I remain committed to continuous improvement, and I am confident that future editions and forthcoming books will benefit greatly from readers' valuable suggestions, guidance, and encouragement.

Dr. Ashok Punde
Pune, India

Preface

Engineering has always been the driving force behind industrial progress. From the earliest machines to the highly sophisticated systems of today, engineering continues to shape the world by transforming ideas into practical solutions.

In the modern industrial environment, the concepts of Ready Engineering, Re-Engineering, and Reverse Engineering have become increasingly important in maintaining, improving, and innovating engineering products and systems.

Among the many specialized areas of engineering, sealing technology plays a critical yet often underappreciated role.

Mechanical seals are essential components in countless rotating machines such as pumps, compressors, mixers, and process equipment used across industries including chemical processing, petrochemicals, power generation, mining, and manufacturing.

A properly designed sealing system ensures operational reliability, safety, environmental protection, and efficiency.

My journey in the field of mechanical sealing began nearly five decades ago, when I started my career as a Mechanical Seal Designer. Over the years, I had the opportunity to work across various sectors of industry, addressing a wide range of sealing challenges in rotating equipment.

Later in my professional career, I served in leadership roles, including Managing Director and Chief Executive Officer, where I spent over twenty-three years guiding engineering teams, developing products, and expanding business operations in the sealing industry.

After retirement from corporate leadership, I continue to remain active in the field as an independent consultant in static and dynamic sealing technologies, focusing on product development, engineering solutions, business enhancement, and mentorship for engineers and organizations.

Throughout these decades of experience, one important realization has consistently stood out: mechanical sealing problems are rarely simple. They require a deep understanding of engineering fundamentals, operating conditions, materials, tribology, manufacturing practices, and equipment behavior.

Many failures in sealing systems arise not because the technology is inadequate, but because the engineering understanding behind the application is incomplete.

This book is an attempt to share the practical engineering philosophy and insights developed over many years of professional experience. ***A central theme is the relationship between three important engineering approaches:***

Ready Engineering — *Designing systems and components from first principles based on defined requirements.*

Re-Engineering — *Improving existing designs and processes to enhance performance, reliability, and efficiency.*

Reverse Engineering — *Analyzing existing products or systems to understand their design, functionality, and limitations.*

Reverse engineering is frequently misunderstood. Many people assume that it simply means copying an existing product. In reality, true reverse engineering is a highly intellectual engineering activity.

It involves careful study of design intent, material behavior, manufacturing techniques, operating conditions, and failure mechanisms. Only with this deep understanding can engineers recreate, improve, or adapt a product effectively.

Therefore, one of the key messages of this book is:

> **Reverse Engineering is not Copy-Paste Engineering.**
>
> It requires knowledge, analytical thinking, creativity, and engineering judgment.

In today's rapidly evolving technological landscape, Artificial Intelligence is beginning to influence many engineering disciplines. AI-assisted simulations, predictive analytics, data-driven maintenance, and intelligent design tools are transforming how engineers approach problem solving and product development.

When combined with traditional engineering expertise, these technologies can significantly enhance the effectiveness of re-engineering and reverse engineering practices.

This book aims to bridge the gap between traditional engineering wisdom and modern technological tools, especially in the context of mechanical sealing systems. It is written primarily for mechanical engineers, design engineers, maintenance engineers, engineering managers, industrial professionals involved in rotating equipment, and young engineers entering the field of sealing technology.

My hope is that this work will help engineers better understand the importance of thoughtful engineering practices, inspire them to approach problems analytically, and encourage them to innovate rather than merely imitate.

Engineering progress does not come from copying what already exists; it comes from understanding, improving, and creating something better.

Dr. Ashok Punde

Independent Consultant — Static & Dynamic Sealing Technology

Expert Appraisal

Andrew Sykes MCGI
Founder
Acumen Seals and Pumps Ltd. United Kingdom.

I have spent 36 years on plant floors and in boardrooms dealing with sealing technology in its most demanding forms. I do not impress easily when it comes to technical literature. So when I tell you that Dr. Ashok Punde has written something genuinely special, I mean it.

What he has achieved here is rare. In nearly four decades in this industry I have rarely come across a book that so effectively bridges the gap between gritty field reality and sound engineering philosophy.

Drawing on 48 years of experience, including over two decades as a Managing Director and CEO, Dr. Ashok Punde has distilled a lifetime of hard-won knowledge into a framework that every maintenance team, every sealing engineer and every technical manager should have on their shelf.

Let me tell you what stands out.

The absolute highlight of this book is Dr. Ashok Punde's treatment of reverse engineering. His central argument is powerful and long overdue: that a purely dimensional copy of a seal will fail if the person doing the copying does not understand the original thermal and hydrodynamic logic behind the design.

He is not anti-reverse engineering. He is deeply pro engineering intelligence. His core philosophy, that reverse engineering without engineering intelligence becomes copying but reverse engineering with engineering intelligence becomes innovation, deserves to become the standard position across our entire industry.

The Engineering Knowledge Ladder he introduces gives young engineers something I wish I had when I was starting out ,a clear vocabulary and a logical progression for how real technical capability is built.

Moving from Ready Engineering through Re-Engineering to true Reverse Engineering is not just a framework. It is a roadmap for developing the kind of intuition that no classroom can teach on its own. What also struck me is his honesty about where failures actually come from. Too many books in this field pin everything on the seal itself.

Dr. Ashok Punde is refreshingly direct about plant floor reality. The majority of mechanical seal failures originate in installation, misalignment and operational factors, not in defective seal design. That is a truth that experienced engineers know but rarely see written down so plainly.

His treatment of Artificial Intelligence is measured and sensible. Rather than following the hype, he positions AI correctly as a powerful data processing tool that cannot replace the physical intuition, engineering judgment and professional responsibility of a human engineer. In a world where AI is being oversold at every turn, that clarity is genuinely valuable.

My final verdict is 9.5 out of 10. This book is timely, practical and grounded in the kind of experience that cannot be faked. Whether you are a young engineer trying to understand why things work the way they do, or a seasoned professional looking for a structured way to mentor your team, Dr. Ashok Punde's contribution is one you will return to. It is a tremendous piece of work, and the mechanical sealing industry is better for it.

Andrew Sykes MCGI,
36 Years Solving Seal Problems,
Founder, Acumen Seals and Pumps Ltd. United Kingdom.

This page has been left blank intentionally.

Professional Perspective

Umar AK Balwa
Managing Director
Sealmatic India Ltd

Dr. Ashok Punde's this book is an essential addition to any rotating equipment engineer's library. Unlike existing literature that often leans too heavily on theory, this work bridges the gap between field experience and modern sealing technology. Dr. Ashok Punde offers profound insights into API plans and advanced monitoring—crucial tools for anyone looking to minimize pump downtime.

Having known Ashok for decades, I have long respected his contributions to the industry. Drawing on 45 years of experience with global leaders like Sealol and AES, he distils complex concepts—such as lubrication films and pressure balances—into actionable knowledge. With clear diagrams and practical case studies, this book is a definitive guide for young engineers and seasoned reliability professionals alike.

A masterclass in Mechanical Seals, Dr. Ashok Punde's book combines many years of frontline expertise with cutting-edge Sealing

Technology. It moves beyond theory to provide practical guidance on API flush plans, material compatibility, and failure analysis.

Whether you are a student or a maintenance technician, this book offers the 'field-tested' insights needed to improve industrial reliability. It is an up-to-date, indispensable resource from a respected veteran who has shaped best practices at various levels of end users, OEMs, EPCs etc.

With Best Wishes. **Umar AK Balwa**

Company information

Sealmatic India Ltd (BSE: 543782) designs and manufactures customized mechanical seals and sealing support systems, catering to oil & gas, refinery, power, petrochemical, marine, aerospace, pulp & paper, fertiliser, pharmaceutical etc. The company headquartered in Mumbai, having three manufacturing plants. The company is known for supplying API-compliant seals to over 63 countries, with significant seals to End Users, EPCs and OE suppliers to KSB, Sulzer, Sundyne, KEPL, Weir, Andritz, KBL, RuhrPumpen, Flowserve, Wilo, SPX, Seepex, Düchting, ITT, BHEL, Idex, Egger, PMSL, MSL, Xylem, Metso, Netzsch, Atlas Copco, Burckhardt and many other renowned OEMS. The company is API Q1 Spec, ISO 19443 (Nuclear), ASME U Stamp, ISO 3834, TRCU 010, 012 & 032, ATEX - 2014/34/EU, DSIR, ISO 9000, 14000, 45000 & PED 97/23/EC QA-System, FDA, GMP, RoHS, REACH Certified Company. *Total Professional Experience: 37 Years*

Endorsement

Pradeep Chapalgaonkar
Chapalgaonkar and Associates, Pune

In the complex and demanding world of industrial machinery, mechanical seals remain critical to reliability and performance—yet their understanding often stays confined to specialists. In this insightful and well-structured work, Dr. Ashok Punde bridges that gap with clarity and depth.

By introducing the practical framework of Ready Engineering, Re-Engineering, and Reverse Engineering, the book offers a fresh and actionable perspective for both design and application engineers. Its strong focus on real-world practices, failure analysis, and evolving challenges—along with reflections on ethics and the growing role of Artificial Intelligence—makes it especially relevant in today's industrial landscape.

Drawing from decades of experience, Dr. Ashok Punde presents complex engineering concepts in a simple, accessible manner. This book is a valuable resource for young engineers and seasoned professionals alike, offering knowledge that is not only informative but immediately applicable.

A Thoughtful, Practical, and Timely Contribution To The Field of Mechanical Seal Technology.

Pradeep Chapalgaonkar
Chapalgaonkar and Associates, Pune- India

Book Review

Dr. Anil Shirahatti
Director (R & D)
Jain Engineers

Mechanical Seals — Ready Engineering, Re-Engineering & Reverse Engineering by Dr. Ashok Punde is a highly valuable and inspiring book, especially from an Indian industrial and entrepreneurial perspective.

In a country like India, where manufacturing, process industries, and engineering self-reliance are becoming increasingly important, this book offers not only technical knowledge but also a practical roadmap for innovation, indigenous development, and engineering excellence.

Before reading this book, mechanical seals may seem like only machine components, but after reading, one gains a much deeper understanding of their importance in process industries from the perspective of industrial reliability, engineering intelligence, and design improvement.

One of the book's greatest strengths is its clear connection between Ready Engineering, Re-Engineering, and Reverse Engineering. This approach is highly relevant to India's growing "Make in India" vision, where understanding existing technologies, improving them intelligently, and building competitive products can strengthen both industry and entrepreneurship.

Dr. Ashok Punde's emphasis that innovation must come from engineering intelligence rather than simple copying is especially meaningful for entrepreneurs who want to build sustainable, credible, and globally competitive businesses.

Dr. Ashok Punde's practical experience and simple explanations make complex engineering concepts easy to understand. His integration of modern tools such as Artificial Intelligence with traditional engineering wisdom makes the book highly relevant for today's engineers, startup founders, and manufacturing leaders.

This Book can positively influence young engineers and industry professionals by encouraging analytical thinking and responsible engineering practices. It is especially recommended for mechanical engineers, maintenance professionals, and those involved in sealing technology or rotating equipment.

Overall, This Book is a valuable guide that combines engineering knowledge, practical wisdom, entrepreneurial vision, and future readiness. It is an inspiring read for anyone who wants to understand engineering beyond theory and contribute meaningfully to India's industrial growth.

Dr. Anil Shirahatti
Director (R & D)

Jain Engineers

Company information

In the complex and demanding world of industrial machinery, mechanical seals remain critical to reliability and performance—yet their understanding often stays confined to specialists. In this insightful and well-structured work, Dr. Ashok Punde bridges that gap with clarity and depth.

Dr. Anil Shirahatti is having 20 years of working knowledge and practical expertise in residual stresses, mechanics of materials, design, product & software development, 3D Printers, Robotics & Automation and high temperature nuclear structural integrity assessment.

About Our Company :

Jain Engineers has been established in the year 1995 with a mission to manufacture world-class products and to provide prompt after-sales services to the clients.

Jain Engineers is a leading Manufacturer & Exporter of Mechanical Shaft Seals under registered trade mark **"JainSeals",** Pump & Distillery Spares, Agitator Drive Assembly, Bearing Isolators, Biotech & Precision Components, Servo hydraulic fatigue Machines, Special Purpose Machines, R & D Test Rigs, 3D Printers etc.

Address :

JAIN ENGINEERS

Plot No. 1224, 12th Main Road KIADB Auto Nagar,

Kanbaragi Industrial Area Belagavi, Karnataka 590 015

GST No. 29ADSPP9887D1ZH

PAN : ADSPP9887D

E-mail : JAINENGINEERS@GMAIL.COM

SALES@JAINSEALS.COM

Mobile No. : 9008501632, 9845096404

This page has been left blank intentionally.

Book Review

Himanshu Bhagwat
Founder Director
Erforschen Techno-Solutions LLP

Dr. Ashok Punde, in this comprehensive work, distills nearly five decades of rich industry experience to illuminate the intricate domain of mechanical sealing technology. The author skillfully guides readers along the "Engineering Knowledge Ladder," progressing from the foundational principles of "Ready Engineering" to the frontiers of industrial innovation.

The book seamlessly balances theory and practice, offering structured methodologies for failure analysis, material selection, and system optimization. It serves not only as a strong conceptual reference but also as a practical, hands-on manual for engineers in the field.

A distinguishing aspect of this work is its emphasis on engineering ethics and responsibility. Dr. Ashok Punde thoughtfully advocates a "Customer-Friendly" design philosophy, encouraging engineers to create solutions that are not only technically sound but also aligned with real-world user needs.

Ultimately, this book stands as a masterclass in transforming extensive industrial experience into a disciplined, structured framework for engineering excellence.

Company Profile

Erforschen Techno-Solutions LLP is a dynamic research and development firm with over four and a half years of experience in delivering innovative engineering solutions. The company specializes in mechanical R&D, with a strong focus on transforming conceptual ideas into practical, high-performance outcomes.

With deep expertise in concept design and engineering validation, Erforschen enables clients to visualize, refine, and systematically develop their ideas at every stage of the innovation cycle. The firm adopts a structured, end-to-end approach—spanning concept development, detailed design, analysis, prototyping, and final implementation—ensuring seamless execution and dependable results.

By combining technical rigor with practical insight, Erforschen Techno-Solutions LLP positions itself as a trusted partner for organizations seeking reliable, efficient, and forward-thinking R&D solutions.

Himanshu Bhagwat
Erforschen Techno-Solutions LLP
Founder Director
Pune – 411038 – India

Table of Contents

PART I

Fundamentals of Engineering and Sealing Technology

PART II

Mechanical Seal Design Engineering

PART VI

Future of Mechanical Sealing Technology

PART I

Fundamentals of Engineering and Sealing Technology

Chapters 1 - 3

This page has been left blank intentionally.

Chapter 1

Engineering Fundamentals and Their Role in Industrial Development

1.1 Introduction

Engineering is the systematic application of scientific knowledge, mathematical principles, and practical experience to design, develop, and improve machines, systems, and industrial processes. It plays a critical role in transforming scientific discoveries into useful technologies that support modern industry and society.

From simple mechanical devices to complex industrial systems, engineering provides the foundation for reliability, efficiency, safety, and productivity. In the modern industrial world, engineering is not limited to designing new equipment; it also includes improving existing systems, solving operational problems, and enhancing performance through innovation.

Mechanical Seals, which are essential components in pumps and rotating equipment, represent an important example of applied engineering. Their design, selection, and performance depend on a deep understanding of engineering fundamentals, including mechanics, materials science, thermodynamics, and tribology.

1.2 Evolution of Engineering Practices

Engineering has evolved significantly over the past century. Earlier industrial systems were often developed through trial-and-error methods and practical experience. While these approaches contributed to many successful inventions, modern engineering increasingly relies on scientific analysis, precise calculations, and advanced design tools. The evolution of engineering practices can broadly be divided into three stages:

Stage 1 — Conventional Engineering

This stage relied heavily on practical experience, manual calculations, and empirical knowledge developed over many years of industrial practice. Engineers carried their knowledge in their hands and in their minds, accumulated through direct observation and problem-solving on the factory floor and in the field.

Stage 2 — Computer-Aided Engineering

With the advancement of computers, engineering design began incorporating simulation, computer-aided design (CAD), and analytical modelling. This greatly improved accuracy and design efficiency. Engineers could now test virtual designs before committing to physical manufacture, reducing development time and costs significantly.

Stage 3 — Intelligent Engineering

THREE STAGES OF ENGINEERING EVOLUTION

In recent years, engineering is entering a new era where artificial intelligence, data analytics, and predictive algorithms assist engineers in decision-making and system optimization. Despite these technological advancements, the fundamental engineering principles remain unchanged. Successful engineering still requires human understanding, judgment, and creativity.

1.3 Role of Engineering in Industrial Reliability

Industrial equipment must operate continuously under demanding conditions such as high pressure, temperature, corrosive environments, and rotating motion. Engineering ensures that machines are designed to withstand these conditions while maintaining safe and reliable operation. One of the key responsibilities of engineering is preventing system failures. Failures in industrial systems can lead to production losses, safety hazards, and environmental risks.

Engineering contributes to reliability through:

- Proper design of mechanical components
- Selection of appropriate materials
- Accurate manufacturing processes
- Correct installation practices

- Continuous monitoring and maintenance

Mechanical sealing systems are particularly critical in maintaining industrial reliability. They prevent leakage of process fluids and ensure the safe operation of pumps, compressors, and other rotating equipment.

1.4 Engineering as a Problem-Solving Discipline

Engineering is fundamentally a problem-solving discipline. Industrial environments constantly present engineers with complex challenges, such as equipment failures, performance limitations, and changing operating conditions. To address these challenges, engineers apply structured approaches that include:

- Problem identification and data collection
- Root cause analysis
- Design modification and testing
- Validation and implementation

This systematic problem-solving approach forms the basis for advanced engineering methodologies such as re-engineering and reverse engineering, which will be discussed in later chapters.

1.5 Engineering Responsibility and Innovation

Engineering is not only about technical knowledge; it also carries significant responsibility. Engineers must ensure that their designs and decisions priorities safety, efficiency, sustainability, and long-term reliability.

Innovation is another key element of engineering. Industrial progress depends on engineers who continuously seek better solutions, improved designs, and more efficient technologies.

In many cases, innovation does not necessarily mean inventing entirely new machines. Often, it involves improving existing systems through deeper analysis, better materials, and smarter design modifications. This is the foundation of both re-engineering and reverse engineering.

1.6 Engineering Foundations for Mechanical Seal Technology

Mechanical seal technology represents a specialized field within mechanical engineering that integrates several engineering disciplines:

- **Tribology** — the study of friction and wear between seal faces
- **Materials Science** — selection of materials capable of withstanding aggressive process conditions
- **Fluid Mechanics** — understanding leakage control and fluid film behaviour
- **Thermodynamics** — managing heat generated at the seal interface
- **Mechanical Design** — ensuring dimensional precision and structural integrity

Because mechanical seals operate under complex conditions involving rotation, pressure, and temperature, their successful design requires strong engineering fundamentals. A weakness in any one of these disciplines will compromise the performance of the entire sealing system.

1.7 Conclusion

Engineering forms the backbone of modern industrial development. Its principles guide the design, operation, and improvement of machines that power industries across the world. In the field of mechanical sealing technology, engineering fundamentals provide the foundation for reliable sealing systems that prevent leakage, enhance equipment life, and support safe industrial operations.

As industries evolve and technologies advance, engineering approaches such as ready engineering, re-engineering, and reverse engineering become increasingly important. The following chapters will explore these concepts in detail, with particular emphasis on their application in mechanical seal engineering in the age of artificial intelligence.

1.8 Engineering Intelligence vs Artificial Intelligence

In recent years, Artificial Intelligence has emerged as a powerful technological tool capable of analyzing large amounts of data, identifying patterns, and assisting engineers in design optimization and predictive maintenance. AI systems can process information much faster than humans and provide valuable insights for improving industrial performance. However, it is important to recognize that Artificial Intelligence does not replace Engineering Intelligence.

Engineering Intelligence is the result of years of education, practical experience, analytical thinking, and problem-solving capability developed by engineers. It includes the ability to understand physical behavior of machines, interpret failure mechanisms, evaluate design limitations, and make informed decisions under complex operating conditions.

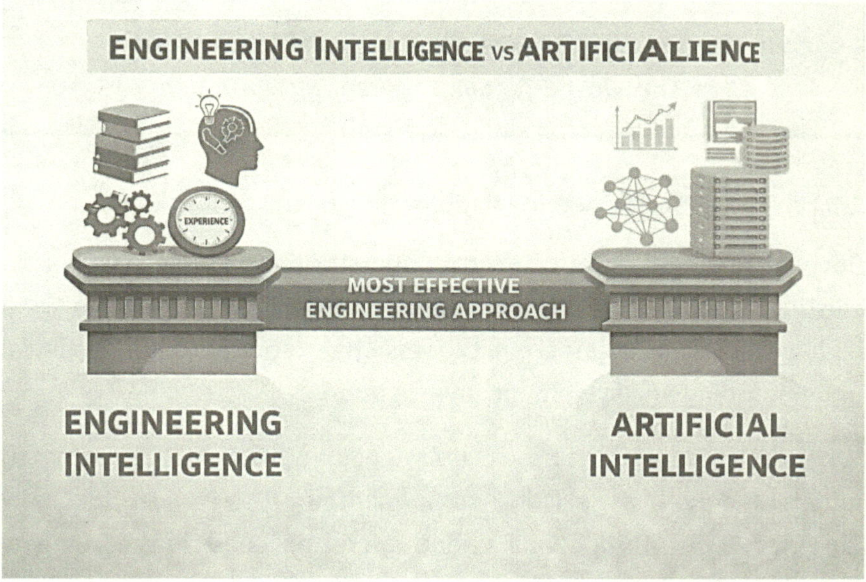

Artificial intelligence systems operate primarily based on data, algorithms, and training models, whereas engineering intelligence is based on scientific understanding, experience, and engineering judgment.

For example, an AI system may analyse vibration data from a pump and predict a potential failure. However, the engineer must interpret the result, identify the root cause, and determine the appropriate corrective action. Without engineering understanding of mechanical systems, the data alone cannot solve the problem.

Artificial intelligence can assist engineers by analyzing operational data, predicting maintenance requirements, supporting design optimization, and monitoring system performance. But the fundamental engineering decisions remain the responsibility of the engineer.

Therefore, in the modern industrial environment, the most effective approach is the integration of Engineering Intelligence with Artificial Intelligence, where AI serves as a tool that enhances human engineering capability rather than replacing it.

Engineering Intelligence + Artificial Intelligence
= The Most Effective Engineering Approach

1.9 Why Reverse Engineering Requires Engineering Knowledge

Reverse engineering is often misunderstood as simply copying an existing product or design. In reality, true reverse engineering is a highly analytical engineering process that requires deep technical understanding and systematic investigation.

When engineers perform reverse engineering, they attempt to understand how an existing component or system was originally designed, how it functions under operating conditions, and why specific materials and geometries were selected.

Without proper engineering knowledge, reverse engineering can easily lead to inaccurate replication, design errors, or reduced reliability. For example, in mechanical seal systems, simply copying the dimensions of an existing seal is rarely sufficient. The original design may have been developed considering factors such as:

- Operating pressure and temperature
- Shaft speed and vibration levels
- Chemical compatibility of materials
- Heat generation at the seal faces
- Equipment alignment and tolerances

If these factors are not fully understood, a copied design may fail prematurely even though its dimensions appear identical to the original. Therefore, successful reverse engineering requires more than measurement — it requires engineering interpretation.

In many industrial situations, reverse engineering also becomes an opportunity for design improvement. Engineers may identify weaknesses in the original design and modify materials, geometry, or operating parameters to enhance performance and reliability.

For this reason, reverse engineering should be viewed not as imitation but as a technical process of understanding, learning, and improving engineering systems.

Chapter 2

Fundamentals of Sealing Technology

2.1 Introduction

In many industrial systems, fluids such as water, chemicals, hydrocarbons, gases, and process liquids are transported, processed, or stored under controlled pressure and temperature conditions.

In such systems, preventing leakage is essential for safe, efficient, and reliable operation. Sealing technology is the branch of engineering that deals with preventing or controlling leakage between two regions that contain fluids under different pressures or environmental conditions.

A properly designed sealing system ensures that the process fluid remains contained within the equipment while preventing external contaminants from entering the system. In modern industries such as chemical processing, petroleum refining, pharmaceuticals, power generation, and water treatment, effective sealing systems play a vital role in maintaining operational integrity and environmental safety.

2.2 Purpose of Sealing in Industrial Equipment

Industrial equipment often contains moving components such as rotating shafts, reciprocating pistons, or sliding surfaces. Whenever two mechanical parts interact while containing a fluid, there exists a potential path for leakage. **The primary objectives of sealing systems include:**

- Preventing leakage of process fluids
- Maintaining pressure within equipment
- Protecting internal components from contamination
- Ensuring safe operation of machinery
- Improving equipment efficiency and reliability

Leakage in industrial systems can lead to several serious consequences, including loss of valuable process fluids, environmental pollution, safety hazards for operating personnel, reduced equipment performance, and increased maintenance and operational costs. Because of these risks, sealing systems must be carefully designed and properly maintained.

2.3 Basic Principles of Sealing

The effectiveness of any sealing system depends on controlling the clearance between two surfaces where leakage could occur. A perfect seal would theoretically allow zero leakage, but in practical engineering applications, sealing systems are designed to reduce leakage to an acceptable and controlled level.

Sealing effectiveness depends on several engineering factors, including surface contact conditions, pressure differences across the seal, fluid properties such as viscosity and temperature, surface finish and dimensional accuracy, and material properties of sealing elements.

In many dynamic sealing systems, a very thin fluid film may exist between the sealing surfaces. This fluid film plays an important role in reducing friction and wear while still maintaining effective leakage control.

2.4 Classification of Sealing Systems

Sealing systems used in industrial equipment can broadly be classified into two major categories.

2.4.1 Static Seals

Static seals are used between two stationary components where there is no relative motion between the sealing surfaces. Examples include gaskets used in flanged pipe connections, O-rings used in stationary housing, and flat seals used in equipment covers. In static sealing applications, the sealing element is compressed between two surfaces to prevent fluid leakage. Although static sealing appears simple, the selection of correct gasket material and proper compression is critical for reliable performance.

2.4.2 Dynamic Seals

Dynamic seals are used where relative motion exists between two components. This motion may be rotational, reciprocating, or oscillating. Examples include mechanical seals in pumps, lip seals used in rotating shafts, and packing seals used in valves and pumps.

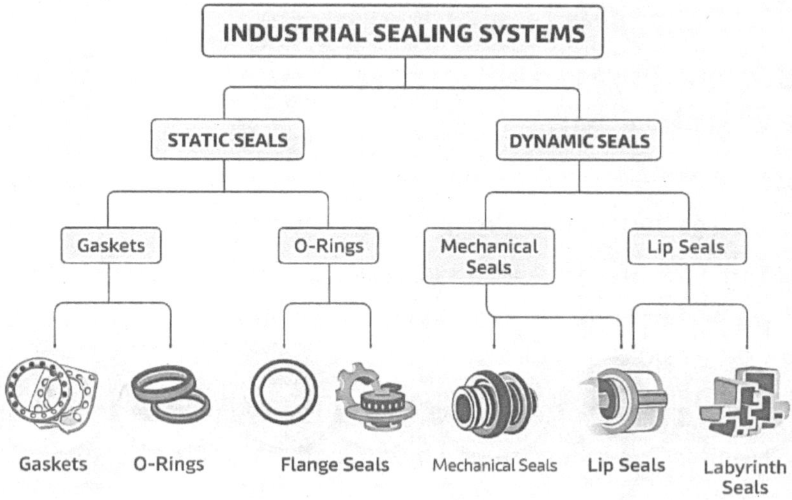

Dynamic sealing applications are more complex than static sealing because the seal must maintain leakage control while accommodating motion, friction, and wear. Mechanical seals represent one of the most advanced and reliable forms of dynamic sealing technology used in modern rotating equipment.

2.5 Leakage Mechanisms in Sealing Systems

Leakage in sealing systems can occur through several mechanisms:

- **Surface leakage** — fluid passes through microscopic gaps between sealing surfaces
- **Material permeation** — certain fluids may slowly penetrate through elastomeric sealing materials
- **Thermal distortion** — temperature changes may cause deformation of sealing components, creating leakage paths
- **Wear and surface damage** — friction between sealing surfaces may gradually increase clearances, leading to leakage

Understanding these leakage mechanisms is essential for engineers involved in the design, maintenance, and troubleshooting of sealing systems.

2.6 Factors Influencing Seal Performance

The performance and reliability of sealing systems depend on several operating parameters:

- **Pressure** — higher pressure differences increase the tendency for leakage and require stronger sealing forces
- **Temperature** — elevated temperatures may affect material properties, causing expansion, degradation, or reduced sealing effectiveness
- **Fluid properties** — viscosity, corrosiveness, and chemical composition influence seal design and material selection
- **Shaft speed** — in rotating equipment, higher speeds increase frictional heat generation and may influence seal stability
- **Alignment and mechanical stability** — misalignment, vibration, and shaft deflection can significantly reduce seal life

These factors must be carefully evaluated when selecting or designing sealing systems for industrial equipment.

2.7 Evolution of Industrial Sealing Methods

Historically, early industrial machinery used simple sealing methods such as soft packing materials placed around rotating shafts. While these methods were relatively simple and inexpensive, they often required frequent adjustment and allowed considerable leakage. The evolution of sealing technology followed the increasing demands of industry:

- **Early period** — rope packing and soft packing seals, requiring regular adjustment and allowing continuous drip leakage
- **Mid period** — improved elastomeric lip seals and compression packing, reducing maintenance frequency
- **Modern period** — precision mechanical seals with engineered face materials, offering near-zero leakage and long service life

Mechanical Seals represented a major technological advancement because they significantly reduced leakage while providing longer operational life compared to traditional packing systems. Today, mechanical seals are widely used in pumps, compressors, mixers, and other rotating equipment across many industries.

2.8 Importance of Sealing Technology in Modern Industry

In modern industrial environments, sealing systems play an increasingly important role due to stricter environmental regulations, higher operating pressures and temperatures, and the need for improved equipment reliability.

Industries now demand sealing solutions that provide minimal leakage, long operational life, reduced maintenance requirements, compatibility with aggressive chemicals, and compliance with environmental safety standards.

2.9 Engineering Perspective of Sealing Systems

From an engineering standpoint, sealing systems should not be viewed merely as small machine components. Instead, they must be considered as critical functional elements of the entire equipment system.

A poorly designed sealing system can compromise the performance of an otherwise well-designed machine. Conversely, a properly engineered sealing system can significantly enhance equipment reliability and operational efficiency.

This perspective is especially important when applying re-engineering or reverse engineering methodologies to sealing problems, as any modification must consider the behavior of the seal within the context of the complete rotating equipment system.

2.10 Conclusion

Sealing technology forms an essential part of industrial machinery and plays a critical role in maintaining safe and efficient operation of equipment that handles fluids under pressure. By preventing leakage, protecting internal components, and maintaining controlled operating conditions, sealing systems contribute directly to industrial productivity and environmental safety.

Among the various sealing methods used in modern industry, mechanical seals represent one of the most advanced and reliable technologies for dynamic sealing applications. The next chapter will introduce the fundamental principles, components, and working mechanisms of mechanical seals.

Chapter 3

Introduction to Mechanical Seals

3.1 Introduction

Mechanical seals are one of the most important sealing devices used in modern rotating equipment. They are widely applied in pumps, compressors, mixers, agitators, and other machinery where a rotating shaft passes through a stationary housing containing process fluid.

The primary function of a mechanical seal is to prevent leakage of the process fluid from the equipment while allowing the shaft to rotate freely.

In many industrial applications, the process fluids may be hazardous, toxic, corrosive, flammable, or environmentally harmful. Therefore, reliable sealing becomes essential not only for equipment efficiency but also for safety, environmental protection, and regulatory compliance.

Mechanical seals were developed to overcome the limitations of traditional shaft packing systems, which often allowed excessive leakage and required frequent maintenance.

3.2 Basic Working Principle of Mechanical Seals

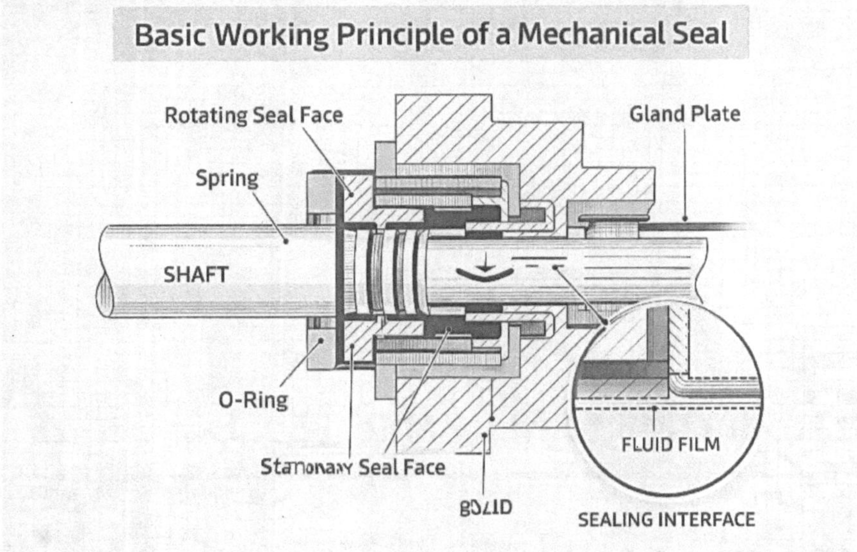

The working principle of a mechanical seal is based on the controlled contact between two very flat sealing surfaces, commonly referred to as seal faces. These two faces include a rotating seal face attached to the shaft and a stationary seal face mounted in the equipment housing.

The seal faces are pressed together by spring force and fluid pressure, creating a sealing interface that restricts the leakage of process fluid.

During operation, the rotating shaft causes one of the seal faces to rotate while the other remains stationary. A very thin fluid film may exist between the faces, which helps reduce friction and wear while still maintaining effective sealing. Because the sealing surfaces are manufactured with extremely high precision and flatness, the leakage path between them is extremely small, resulting in minimal and controlled leakage.

3.3 Major Components of a Mechanical Seal

A typical mechanical seal consists of several important components that work together to provide effective sealing.

3.3.1 Rotating Seal Face

The rotating seal face is mounted on the rotating shaft and rotates along with it. This component forms one half of the primary sealing interface. It is usually manufactured from materials with high wear resistance and good thermal properties. The quality of its surface finish is critical — flatness to within fractions of a micron is typically required.

3.3.2 Stationary Seal Face

The stationary seal face is fixed within the seal housing or gland plate. It forms the other half of the sealing interface. The stationary face must maintain precise alignment with the rotating face to ensure uniform contact and proper sealing performance throughout the operational life of the seal.

3.3.3 Secondary Sealing Elements

Secondary seals prevent leakage between the seal components and the equipment surfaces. Common secondary sealing elements include O-rings, gaskets, and elastomeric bellows. These components ensure that the process fluid does not bypass the primary sealing faces. Their material selection is critically important, as they must be chemically compatible with the process fluid and thermally stable at operating temperatures.

3.3.4 Springs or Bellows

Springs provide the necessary axial force to keep the seal faces in contact during operation. They compensate for face wear, shaft movement, thermal expansion, and minor misalignment. In some designs, metal bellows or elastomeric bellows are used instead of conventional springs. The spring force must be carefully calculated —

too little force results in leakage, while excessive force causes rapid face wear and heat generation.

3.3.5 Metal Components

Mechanical seals also include various metallic parts such as seal glands, drive collars, and retainers. These components provide structural support and ensure proper positioning of the seal assembly. Material selection for these parts must consider corrosion resistance, mechanical strength, and compatibility with the process environment.

3.4 Primary and Secondary Sealing Functions

Mechanical seals operate using two levels of sealing. Primary sealing occurs at the interface between the rotating and stationary seal faces. This is the main sealing point that controls leakage of the process fluid. Secondary sealing occurs at various locations within the seal assembly to prevent fluid from bypassing the primary seal faces. These include O-rings and gaskets placed between different seal components. Both sealing functions are essential for maintaining reliable seal performance.

3.5 Seal Face Materials — Overview

The performance and life of a mechanical seal depend heavily on the materials used for the seal faces. Common seal face materials include carbon graphite, silicon carbide, tungsten carbide, and ceramic materials.

These materials are selected based on wear resistance, thermal conductivity, chemical compatibility, and friction characteristics. The proper combination of seal face materials is critical for achieving long seal life and stable operation. **A detailed discussion of materials is provided in Chapter 6.**

3.6 Advantages of Mechanical Seals Over Traditional Packing

Mechanical seals offer several advantages compared with conventional packing systems:

- **Significantly reduced leakage** — near-zero emission under normal operating conditions
- **Longer service life** — properly designed seals can operate for years without replacement
- **Lower maintenance requirements** — reduced frequency of adjustment and intervention
- **Improved equipment efficiency** — less friction loss compared to packed glands
- **Reduced environmental contamination** — critical for hazardous or toxic process fluids
- **Compliance with environmental regulations** — increasingly mandatory in process industries

3.7 Basic Types of Mechanical Seals

Mechanical Seals can be classified into several types depending on their design and operating characteristics:

- **Single mechanical seals** — used in relatively simple applications where leakage control requirements are moderate and the pumped fluid can act as a lubricant
- **Double mechanical seals** — used in hazardous or high-pressure applications where additional sealing protection is required; a barrier fluid is used between the two seals
- **Balanced and unbalanced seals** — classification based on pressure distribution across the seal faces; balanced seals are used for higher pressure applications
- **Cartridge seals** — pre-assembled mechanical seals designed for easier installation and improved reliability; they eliminate most installation errors

- **Pusher seals** — use axial spring force to maintain face contact, suitable for a wide range of applications
- **Non-pusher seals** — use bellows instead of springs; eliminate the problem of chemical attack on spring materials

3.8 Operating Conditions Affecting Mechanical Seals

Mechanical seals must operate under a wide range of conditions that influence their performance. Important operating parameters include:

- **Fluid pressure** — determines the hydraulic forces acting on the seal faces
- **Operating temperature** — affects material properties and lubrication behaviour
- **Shaft speed** — directly influences heat generation and face stability
- **Fluid properties** — viscosity, chemical composition, presence of solids
- **Equipment alignment and vibration** — affects the uniformity of face contact

Engineers must carefully evaluate all these factors when selecting or designing mechanical seals for specific applications. Overlooking even one parameter can result in premature seal failure.

3.9 Engineering Importance of Mechanical Seals

Mechanical seals may appear to be relatively small components in large industrial machines, but their importance is significant. A failure of a mechanical seal can lead to leakage of hazardous fluids, equipment shutdown, production losses, environmental damage, and safety risks.

Therefore, mechanical seal design, selection, and maintenance require careful engineering attention.

In many industries, the reliability of pumps and rotating equipment depends heavily on the performance of the mechanical sealing system. The cost of a single unplanned shutdown caused by a seal failure often exceeds the cost of several years of preventive maintenance.

This economic reality underscores the critical importance of engineering intelligence in seal selection and design.

3.10 Conclusion

Mechanical seals represent a critical technology in modern rotating equipment. By providing effective leakage control and reliable operation under demanding conditions, they contribute significantly to industrial productivity and safety.

Understanding the fundamental principles, components, and operating characteristics of mechanical seals is essential for engineers involved in the design, operation, and maintenance of industrial machinery.

The next chapter will explore Ready Engineering in Mechanical Seal Design, where we will discuss how engineers design sealing systems by considering operating conditions, equipment characteristics, and engineering fundamentals.

This page has been left blank intentionally.

PART II

Mechanical Seal Design Engineering

Chapters 4 — 6

This page has been left blank intentionally.

Chapter 4

Ready Engineering in Mechanical Seal Design

4.1 Introduction

Engineering design in industrial equipment can follow different approaches depending on the availability of knowledge, experience, and existing technologies. In the field of mechanical sealing technology, engineers often need to develop reliable sealing solutions within limited time while ensuring safe and efficient operation.

One practical approach widely used in industry is Ready Engineering.

Ready Engineering refers to an engineering methodology in which new designs are developed by combining fundamental engineering principles with proven design knowledge and established industrial practices.

This approach enables engineers to create reliable solutions efficiently without starting every design entirely from the beginning. By integrating scientific understanding with practical experience, Ready Engineering provides a balanced and effective method for mechanical seal design.

4.2 Concept of Ready Engineering

Ready Engineering can be understood as a structured design approach that includes two essential and complementary elements:

1. **Application of engineering fundamentals and scientific principles**
2. **Use of existing design knowledge, field experience, and proven engineering solutions**

In this approach, engineers do not simply copy existing designs. Instead, they analyse known designs, understand the engineering logic behind them, and adapt them to meet the requirements of a new application.

For example, when designing a mechanical seal for a pump operating under specific pressure, temperature, and speed conditions, engineers may begin with an established seal configuration that has been successfully used in similar applications. They then modify materials, dimensions, and design parameters to match the operating conditions of the new system.

Thus, Ready Engineering represents engineering with informed preparation, where the engineer uses available knowledge as a starting point while applying analytical reasoning to create a reliable design. This distinguishes it from simple replication and elevates it to a genuine engineering discipline.

4.3 Ready Engineering in Mechanical Seal Applications

Mechanical seals operate in complex environments involving rotating motion, fluid pressure, temperature variations, and chemical exposure. Designing a mechanical seal requires careful evaluation of these operating conditions. Typical Ready Engineering activities in mechanical seal design include:

- Selecting a suitable seal configuration based on equipment type and application

- Choosing appropriate seal face material combinations
- Determining spring arrangements and the correct sealing force
- Selecting elastomers chemically compatible with the process fluid
- Adapting seal dimensions to shaft and housing geometry
- Evaluating thermal conditions and specifying cooling or flushing arrangements

Through this method, engineers achieve reliable designs while reducing development time and minimizing risk. The process is efficient because it builds on the accumulated knowledge of the industry, while still requiring the engineer to exercise genuine technical judgment.

4.4 Role of Engineering Fundamentals

Although Ready Engineering benefits from existing design knowledge, it must always be supported by fundamental engineering principles. Key engineering disciplines involved in mechanical seal design include:

- **Tribology** — understanding friction, wear, and lubrication behaviour between seal faces
- **Fluid Mechanics** — analyzing pressure distribution and leakage paths within the sealing interface
- **Thermodynamics** — evaluating heat generation and heat dissipation at the seal faces
- **Materials Engineering** — selecting materials capable of withstanding mechanical stress, chemical exposure, and temperature variations

Without a proper understanding of these engineering fundamentals, the adaptation of existing designs may lead to unreliable or unsafe results. Ready Engineering is not a shortcut; it is an informed and disciplined starting point for competent seal design.

4.5 Advantages of Ready Engineering

Ready Engineering offers several important advantages in industrial design practice:

- Efficiency in design development — using established design frameworks reduces the time required to develop reliable sealing solutions
- **Reduced risk of failure** — proven engineering concepts provide a foundation of reliability based on previous successful applications
- **Practical adaptability** — existing designs can be modified to suit new operating conditions without unnecessary complexity
- **Integration of experience and knowledge** — Ready Engineering allows engineers to combine theoretical knowledge with practical field experience

4.6 Limitations and Engineering Responsibility

While Ready Engineering is highly effective, it must be applied with proper engineering judgment. Simply selecting an existing design without analyzing its suitability for the new application can lead to performance problems.

Engineers must carefully evaluate operating pressure and temperature, chemical compatibility of materials, equipment alignment and vibration conditions, shaft speed and mechanical stability, and process safety requirements.

If these factors are not considered, even a well-proven design may fail under different operating conditions. Ready Engineering requires responsible engineering analysis rather than simple design reuse.

4.7 Ready Engineering as a Foundation for Re-Engineering and Reverse Engineering

Ready Engineering represents the first stage in a broader engineering framework that includes re-engineering and reverse engineering. Ready Engineering focuses on developing new designs using engineering principles and established knowledge.

Re-Engineering involves improving or modifying existing designs to enhance performance, reliability, or efficiency. Reverse Engineering focuses on analyzing existing components or systems to understand their design and function.

Together, these three engineering approaches form a comprehensive methodology for solving complex industrial problems. The relationship between these three approaches is best understood through the Engineering Knowledge Flow concept introduced in Section 4.9.

4.8 Conclusion

Ready Engineering provides a practical and efficient approach to mechanical seal design by combining fundamental engineering principles with proven industrial knowledge. Rather than relying solely on theoretical design or simple replication of existing products, Ready Engineering encourages engineers to analyse, adapt, and apply engineering knowledge intelligently.

This methodology allows engineers to develop reliable sealing solutions while maintaining efficiency in design development and industrial implementation.

4.9 Engineering Knowledge Flow

Engineering development in industrial technology often follows a progressive flow of knowledge and experience. This progression can be understood through the concept of Engineering Knowledge Flow, which connects different engineering methodologies used in design, improvement, and innovation.

Engineering Knowledge Flow in Mechanical Seal Development

Ready Engineering represents the starting point of practical engineering design. In this stage, engineers develop new systems or components using engineering fundamentals together with established design knowledge and proven industrial practices.

Re-Engineering focuses on improving existing systems, where engineers analyse current designs and introduce modifications to enhance efficiency, reliability, durability, or cost effectiveness.

Engineering Knowledge Flow Diagram

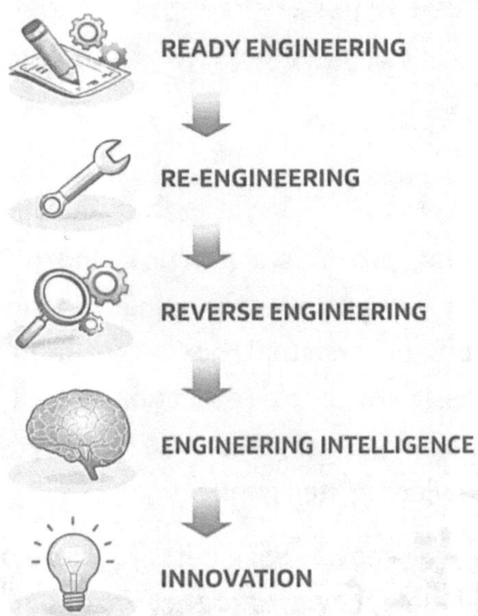

Reverse Engineering involves systematic analysis of existing products to understand their design, materials, and functional principles. Innovation represents the ultimate outcome of accumulated engineering knowledge — when engineers integrate insights gained from all earlier stages, they develop new and advanced technologies.

In the field of mechanical sealing systems, this knowledge flow plays a significant role in technological progress. Engineers first develop sealing systems through Ready Engineering.

Operational experience leads to Re-Engineering improvements. Detailed analysis of existing designs may involve Reverse Engineering. The integration of these insights ultimately results in Innovation in sealing technology.

This continuous flow of engineering knowledge drives the evolution of more reliable and efficient mechanical sealing systems.

This page has been left blank intentionally.

Chapter 5

Tribology and Heat Management in Mechanical Seals

5.1 Introduction

The reliable operation of mechanical seals depends greatly on the behavior of the contacting surfaces that form the sealing interface. These surfaces are subject to friction, wear, lubrication, and heat generation during operation.

The scientific study of these phenomena is known as tribology. Tribology is a multidisciplinary field that combines mechanical engineering, materials science, and surface physics to understand how surfaces interact when they move relative to each other.

In mechanical seals, the two seal faces operate under controlled contact conditions while one face rotates relative to the other. The behavior of this interface determines seal leakage, frictional losses, heat generation, wear rate, and ultimately seal life.

If friction and heat generation are not properly controlled, the seal faces may experience excessive wear, distortion, or thermal damage, leading to seal failure. Understanding tribological behavior and implementing proper heat management strategies are therefore essential for successful mechanical seal design and operation.

5.2 Fundamentals of Tribology

Tribology focuses on three primary phenomena — friction, wear, and lubrication. These three elements interact closely and determine the performance of the sealing interface. In mechanical seals, the goal is to maintain a controlled balance between friction and lubrication so that the seal faces remain stable, with minimal wear and minimal leakage.

5.2.1 Friction

Friction is the resistance to motion that occurs when two surfaces slide against each other. In mechanical seals, friction occurs between the rotating seal face and the stationary seal face.

The frictional force depends on contact pressure between seal faces, surface roughness, material properties, presence of lubrication, and sliding speed. Friction cannot be completely eliminated in mechanical seals because some level of contact is required to maintain effective sealing.

However, excessive friction can lead to heat generation, surface damage, and accelerated wear.

5.2.2 Wear

Wear is the gradual removal or deformation of material from surfaces in contact due to mechanical interaction. Common wear mechanisms in mechanical seals include:

- **Adhesive wear** — occurs when microscopic bonding takes place between contacting surfaces and fragments of material are removed during sliding motion
- **Abrasive wear** — occurs when hard particles or rough surface asperities scratch or cut into the seal faces

- **Corrosive wear** — occurs when chemical reactions between the process fluid and seal materials weaken the surfaces
- **Thermal wear** — occurs when excessive heat causes surface degradation or distortion

Proper material selection and lubrication conditions are critical to reducing wear. Understanding the dominant wear mechanism in a specific application is essential for selecting the correct seal face material combination.

5.2.3 Lubrication

Lubrication plays a crucial role in reducing friction and wear between seal faces. In mechanical seals, lubrication is often provided by a thin fluid film formed by the process of the fluid itself. This thin film separates the seal faces partially while still maintaining a controlled sealing interface.

The presence of this fluid film reduces friction between seal faces, minimizes wear, removes heat generated at the interface, and stabilizes seal operation. Maintaining a stable lubrication film is essential for long seal life.

5.3 Seal Face Lubrication Regimes

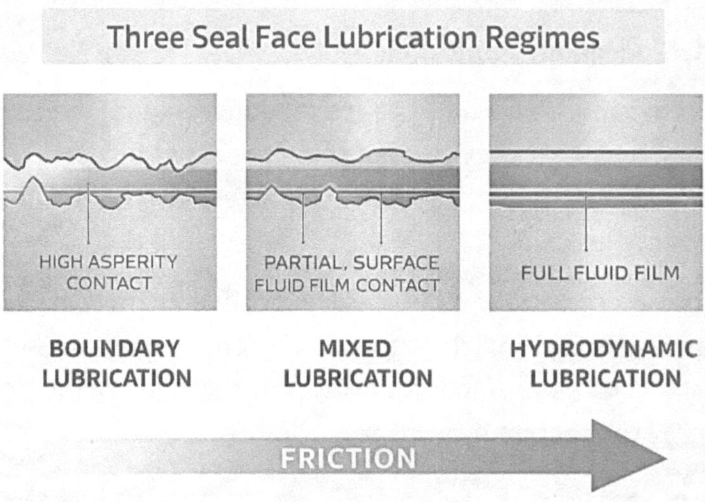

The interaction between the seal faces can operate under different lubrication conditions depending on operating parameters such as speed, load, and fluid properties. These conditions are categorized into three lubrication regimes.

5.3.1 Boundary Lubrication

In boundary lubrication, the fluid film between the seal faces is extremely thin, and significant surface contact occurs between the faces. Under this condition, friction levels are relatively high, wear is more likely to occur, and surface material properties become very important.

Boundary lubrication may occur during start-up, shutdown, or low-speed operation. This is the most critical period in a seal's life, and material selection must specifically address this regime.

5.3.2 Mixed Lubrication

In mixed lubrication, both surface contact and fluid film lubrication occur simultaneously. Some areas of the seal faces are separated by fluid, while others experience direct contact. This regime is commonly encountered during normal mechanical seal operation. The proportion of contact to fluid-film areas determines the effective friction coefficient and wear rate.

5.3.3 Hydrodynamic Lubrication

In hydrodynamic lubrication, a continuous fluid film completely separates the seal faces. Under this condition, friction is greatly reduced and wear is minimal. However, excessive fluid film thickness may increase leakage.

Therefore, mechanical seals are designed to maintain a controlled thin film rather than complete separation. Some advanced seal face designs incorporate spiral grooves or other geometric features specifically to generate hydrodynamic lift.

5.4 Heat Generation in Mechanical Seals

One of the most important consequences of friction at the seal faces is heat generation. As the rotating face slides against the stationary face, friction converts mechanical energy into heat.

This heat must be effectively managed to prevent damage to the seal components. Excessive heat can cause thermal distortion of seal faces, degradation of elastomeric components, loss of lubrication, increased wear, and ultimately seal failure.

Understanding the mechanisms of heat generation and dissipation is therefore essential in seal design.

5.5 Frictional Heat Generation — Basic Engineering Relation

The amount of heat generated at the seal interface depends primarily on the friction force and the sliding velocity.

A simplified engineering expression for frictional heat generation is:

Heat Generated (Q) = Friction Force × Sliding Velocity

$$Q = \mu \times p \times A \times v$$

where μ = coefficient of friction, p = contact pressure, A = effective face area, v = sliding velocity

As rotational speed increases, frictional heat generation increases significantly. This is why high-speed applications require special attention to cooling arrangements, material selection with high thermal conductivity, and face geometry designed to minimise friction while maintaining effective sealing.

5.6 Temperature Effects on Seal Performance

Temperature rise at the seal faces can influence several aspects of seal operation:

- **Thermal Distortion** — seal face materials expand when heated; uneven temperature distribution may cause thermal distortion, leading to uneven contact and increased leakage
- **Loss of Lubrication** — high temperatures may cause evaporation or degradation of the lubricating fluid film
- **Elastomer Degradation** — secondary sealing elements such as O-rings may lose elasticity at high temperatures
- **Material Degradation** — certain face materials may lose strength or hardness when exposed to excessive heat

Proper thermal control is therefore essential for reliable seal operation. The thermal behavior of the sealing system must be evaluated at the design stage, not left to chance.

5.7 Heat Dissipation Methods in Mechanical Seals

To maintain acceptable operating temperatures, mechanical seals are designed with various heat dissipation mechanisms:

- **Fluid Circulation** — the process fluid itself often acts as a cooling medium by circulating around the seal faces and carrying away heat
- **External Flush Systems** — in some applications, external fluid is introduced near the seal faces to remove heat and provide lubrication; flush systems are commonly used in high-temperature or high-pressure applications
- **Seal Chamber Design** — the geometry of the seal chamber influences fluid circulation and heat removal; proper chamber design can significantly improve cooling efficiency

- **Heat-Conductive Materials** — materials such as silicon carbide and tungsten carbide have high thermal conductivity and can dissipate heat effectively
- **Cooling Jackets** — some seal housings include cooling jackets through which water or other cooling fluids circulate to control seal temperature

The selection of the appropriate heat dissipation method depends on the specific application, operating conditions, and the severity of the thermal environment. API standards define various flush plans (such as Plan 11, Plan 23, Plan 32, and others) that standardize cooling arrangements for different application conditions.

5.8 Engineering Control of Seal Face Temperature

Engineers must carefully control several parameters to maintain stable seal face temperature. Important control factors include selection of appropriate seal face materials, proper contact pressure between faces, adequate lubrication conditions, effective cooling methods, and proper alignment of rotating components.

If these factors are properly managed, mechanical seals can operate reliably for long periods without excessive wear or overheating.

5.9 Tribology in Failure Analysis

Tribological analysis is often used in investigating mechanical seal failures. Common tribological failure indicators include scoring marks on seal faces, uneven wear patterns, thermal cracks, and surface discoloration due to overheating.

By studying these patterns, engineers can identify the root cause of seal failure and implement corrective actions. This type of analysis also plays an important role in re-engineering and reverse engineering processes, discussed in later chapters.

5.10 The Mechanical Seal Engineering Triangle

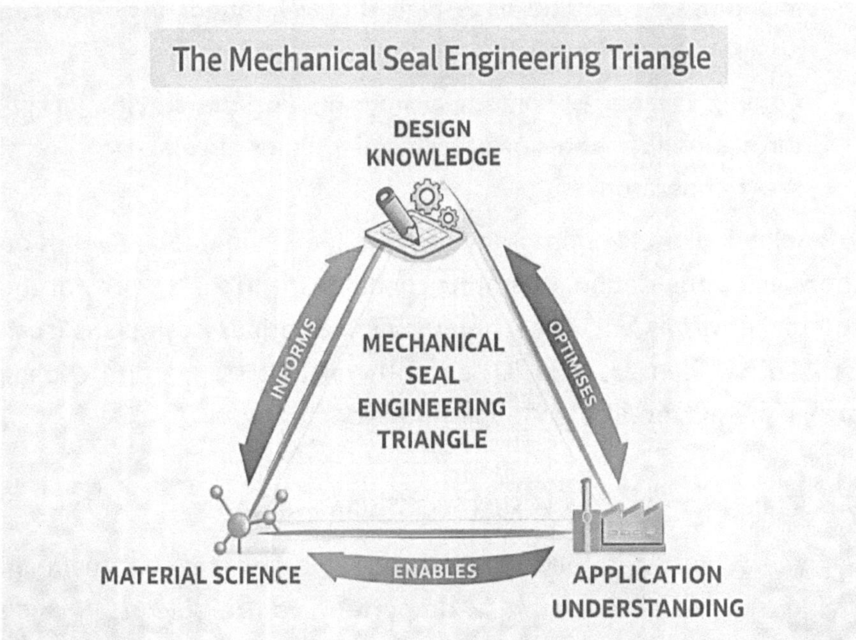

Successful mechanical seal engineering requires the balanced integration of three major knowledge domains.

A weakness in any one area will compromise the reliability of the entire sealing system.

- **Design Knowledge** — seal face geometry, spring loading, pressure balance, secondary seals, seal chamber configuration
- **Material Science** — face materials, elastomers, metal alloys, chemical compatibility, thermal properties
- **Application Understanding** — process fluid characteristics, operating pressure and temperature, shaft speed, equipment condition, vibration levels

Because these three elements are strongly interconnected, mechanical seal engineering requires a holistic approach. A change in application conditions may require a change in materials.

A change in materials may require modifications in design. A change in design geometry affects heat generation and therefore influences material selection. Experienced engineers gradually learn to navigate these interdependencies — this is the essence of engineering intelligence in sealing technology.

5.11 Conclusion

Tribology and heat management are fundamental aspects of mechanical seal performance. The interaction between seal faces involves complex relationships between friction, wear, lubrication, and heat generation.

By carefully controlling these factors through proper design, material selection, and cooling methods, engineers can ensure stable seal operation and long service life. Understanding tribological behavior also provides valuable insights into seal failure mechanisms, which supports engineering activities such as troubleshooting, re-engineering, and reverse engineering. The next chapter will discuss material selection in mechanical seals.

This page has been left blank intentionally.

Chapter 6

Material Selection in Mechanical Seals

6.1 Introduction

Material selection is one of the most critical aspects of mechanical seal design. The performance, reliability, and service life of a mechanical seal depend greatly on the materials used in its various components.

Mechanical seals operate under challenging conditions that may involve high rotational speeds, elevated temperatures, high pressures, chemical exposure, and abrasive particles in fluids. Under these conditions, improper material selection can lead to rapid wear, corrosion, deformation, or seal failure.

Engineers must carefully evaluate the operating environment and select materials that provide the required mechanical strength, chemical resistance, thermal stability, and tribological performance.

6.2 Seal Face Materials

The seal faces form the primary sealing interface of a mechanical seal. Because these surfaces slide against each other during operation, their material properties must be carefully selected to ensure low friction, high wear resistance, good thermal conductivity, and chemical compatibility with process fluids.

6.2.1 Carbon Graphite

Carbon graphite is one of the most widely used materials in mechanical seal applications. Its excellent self-lubricating properties arise from the layered crystal structure of graphite, which allows easy inter-layer sliding. Key characteristics include:

- **Excellent self-lubricating properties** — can operate with minimal external lubrication
- **Low coefficient of friction** — reduces heat generation at the sealing interface
- **Good thermal stability** — maintains properties across a wide temperature range
- **Ability to operate under boundary lubrication conditions** — critical during start-up and shutdown

Because of these properties, carbon graphite is commonly used as the softer mating face in many seal combinations. However, carbon materials have relatively lower mechanical

strength compared to carbide materials and may not be suitable for extremely abrasive environments or applications involving strong oxidising chemicals.

6.2.2 Silicon Carbide

Silicon carbide (SiC) is a highly advanced ceramic material widely used in modern mechanical seals. Its combination of hardness, thermal conductivity, and chemical resistance makes it one of the most versatile seal face materials available. Key properties include:

- **Very high hardness** — excellent resistance to abrasive wear
- **High thermal conductivity** — effectively dissipates frictional heat

- **Outstanding chemical resistance** — suitable for acids, alkalis, and organic solvents
- **Ability to withstand high temperatures** — maintains properties at elevated operating temperatures

Silicon carbide is available in several grades, including pressureless sintered SiC and reaction-bonded SiC, each with slightly different properties suited to different applications. It is often paired with carbon graphite to form a reliable and thermally stable seal face combination.

6.2.3 Tungsten Carbide

Tungsten carbide (WC) is another high-performance material used in mechanical seals, particularly in demanding industrial applications. Key characteristics include:

- **High mechanical strength and toughness** — resistant to cracking under shock loads
- **Excellent resistance to abrasion** — ideal for slurry and particle-laden fluid applications
- **High resistance to deformation** — maintains dimensional stability under load
- **Good thermal conductivity** — contributes to heat dissipation

Tungsten carbide surfaces can generate higher friction compared with some other material combinations. This must be carefully considered in seal design, particularly through proper face balance and flushing arrangements. Different binder contents (typically cobalt or nickel) affect corrosion resistance.

6.2.4 Ceramic Materials

Alumina ceramic (Al2O3) is sometimes used in mechanical seals for specific applications, particularly in water and food processing industries. Typical properties include high hardness, good corrosion resistance, and good dimensional stability.

However, ceramics are generally more brittle than carbide materials and may be sensitive to mechanical shock or rapid thermal cycling. Their use has diminished as silicon carbide has become more widely available and cost-effective.

6.3 Common Seal Face Material Combinations

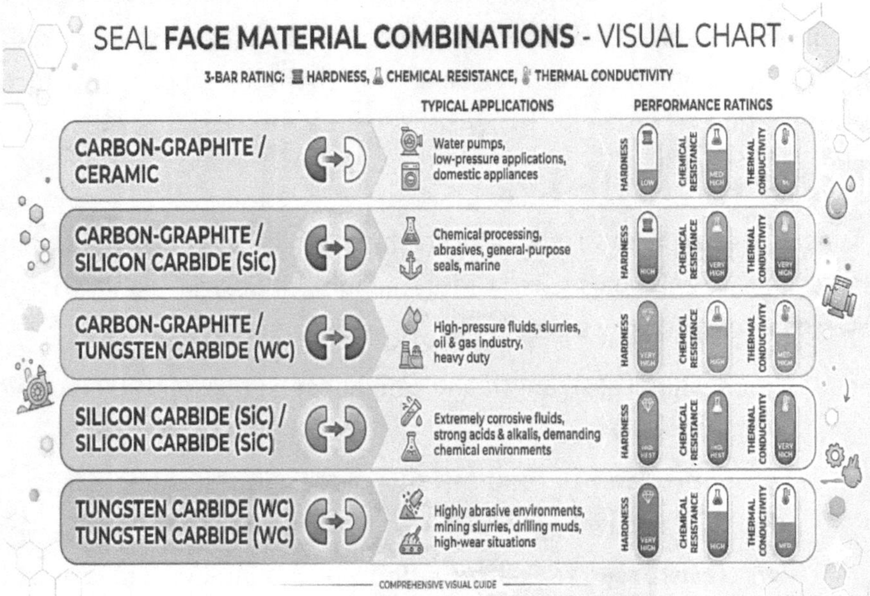

The effectiveness of a mechanical seal depends not only on individual materials but also on how the materials interact with each other. Some common seal face combinations and their typical applications are summarized below:

Rotating Face	Stationary Face	Typical Application
Carbon Graphite	Alumina Ceramic	General industrial pumps, clean water service
Carbon Graphite	Silicon Carbide	Chemical processing, hydrocarbon service
Carbon Graphite	Tungsten Carbide	Light abrasive environments, petroleum products
Silicon Carbide	Silicon Carbide	High temperature, corrosive fluids, slurry service

Tungsten Carbide	Tungsten Carbide	Heavy abrasive slurry applications

6.4 Secondary Sealing Materials — Elastomers

Secondary seals prevent leakage around the mechanical seal components and provide flexibility for movement and thermal expansion. These sealing elements are commonly made from elastomeric materials selected based on chemical compatibility and temperature resistance.

6.4.1 Nitrile Rubber (NBR)

Nitrile rubber is widely used in mechanical seals due to its good resistance to oils and petroleum products. It offers good mechanical strength and moderate temperature resistance (typically up to approximately 100°C). It is the most economical elastomer for general service. However, nitrile rubber may not be suitable for strong chemicals, aromatic hydrocarbons, or very high temperatures.

6.4.2 Fluoroelastomer (FKM / Viton)

Fluoroelastomers are widely used in chemical and high-temperature applications. They offer excellent chemical resistance across a broad range of acids, alkalis, and solvents, along with high temperature tolerance (typically up to 200°C or higher in certain grades).

Viton is the preferred elastomer in chemical processing and refinery applications where standard nitrile rubber would degrade rapidly.

6.4.3 EPDM (Ethylene Propylene Rubber)

EPDM elastomers provide excellent resistance to water, steam, and certain non-petroleum-based chemicals. They are commonly used in water treatment systems, steam applications, and certain chemical processes. However, EPDM is generally not compatible with petroleum-based fluids, which would cause rapid swelling and degradation.

6.4.4 PTFE (Polytetrafluoroethylene)

PTFE is used in aggressive chemical environments where elastomers may degrade. It offers exceptional chemical resistance across virtually the entire range of industrial chemicals, a wide usable temperature range, and low friction properties. Because PTFE has limited elasticity, it is used in special designs that compensate for its lower flexibility, such as PTFE wedge secondary seals or PTFE-encapsulated O-rings.

6.5 Materials for Springs and Metal Components

Mechanical seals include several metallic parts such as springs, retainers, and seal glands. Common materials used for these components include:

- **Stainless steel (316SS)** — standard material for most industrial applications; good corrosion resistance in many environments
- **Hastelloy C-276** — for highly corrosive environments including strong acids and chloride-containing media
- **Titanium** — for specific aggressive chemical environments including oxidising acids
- **Alloy 20** — for sulphuric acid and other aggressive environments

The selection of metallic materials depends on the corrosive nature of the process fluid, mechanical load requirements, and long-term fatigue resistance, particularly for spring components that undergo repeated cycling.

6.6 Chemical Compatibility Considerations

Chemical compatibility is one of the most important factors in selecting seal materials. Process fluids may contain chemicals that corrode metals, degrade elastomers, or cause swelling or hardening of materials.

Engineers must carefully evaluate the chemical composition of the fluid and ensure that all seal materials are compatible with the process environment. Comprehensive material

compatibility charts are available as a reference guide, but laboratory testing is always recommended for critical or unusual chemical services.

A small incompatibility can lead to catastrophic early failure.

6.7 Material Failures in Mechanical Seals

Improper material selection can lead to several types of failure:

- **Chemical corrosion** — attack of metals or elastomers by aggressive chemicals, causing dimensional changes or loss of material integrity
- **Thermal degradation** — loss of material properties due to excessive heat, including softening of metals, hardening of elastomers, or oxidation of carbon faces
- **Mechanical fracture** — cracking or breakage caused by stress, impact, or thermal shock, particularly in brittle ceramics or carbides
- **Elastomer swelling or hardening** — changes in elastomer dimensions and properties due to chemical absorption, causing leakage or loss of flexibility

6.8 Engineering Considerations in Material Selection

Material selection should not be based on a single property. Engineers must evaluate multiple factors simultaneously, including mechanical strength, wear resistance, thermal properties, chemical compatibility, and cost and availability.

Often the best material combination is achieved through engineering balance rather than maximizing one property alone. For example, silicon carbide may offer superior hardness, but if it is paired with another equally hard face in a low-lubrication environment, the result may be catastrophic brittle fracture rather than reliable operation.

6.9 Importance of Material Engineering in Mechanical Seals

Material science plays a fundamental role in modern mechanical seal technology. Advances in ceramic materials, carbides, and high-performance elastomers have significantly improved seal performance and durability.

These advancements have enabled mechanical seals to operate under more demanding conditions than ever before — in high-pressure chemical processes, high-temperature applications, highly corrosive environments, and abrasive slurry services.

Material engineering therefore continues to be a key driver of innovation in sealing technology.

6.10 Conclusion

Material selection is a critical element of mechanical seal design. Proper choice of seal face materials, elastomers, and metallic components ensures reliable performance under challenging industrial conditions.

Engineers must carefully consider the operating environment, chemical compatibility, and mechanical requirements when selecting materials for sealing systems. By combining appropriate materials with sound engineering design, mechanical seals can achieve long service life, minimal leakage, and reliable operation.

The next chapter will focus on the concept and importance of re-engineering.

PART III

Re-Engineering of Mechanical Sealing Systems

Chapters 7 — 8

This page has been left blank intentionally.

Chapter 7

Concept and Importance of Re-Engineering

7.1 Introduction

Industrial systems continuously evolve due to changing operational requirements, technological advancements, and the need for improved efficiency and reliability. As equipment operates over time, engineers gain valuable insights into its performance, limitations, and areas for improvement.

In many situations, existing designs may not fully meet new operational demands or may exhibit performance issues such as excessive wear, leakage, or frequent maintenance requirements. Instead of completely replacing the equipment or developing entirely new designs, engineers often apply a systematic process known as Re-Engineering.

Re-Engineering refers to the process of analyzing existing systems or components and modifying their design, materials, or operating conditions to improve performance, reliability, or efficiency. In mechanical seal technology, re-engineering plays a crucial role in enhancing the performance of sealing systems across many industries.

7.2 Definition of Re-Engineering

Re-Engineering can be defined as the systematic process of analyzing an existing engineering design and introducing modifications to improve its performance, reliability, efficiency, or suitability for new operating conditions.

Unlike reverse engineering, which focuses on understanding the design of an existing product, re-engineering focuses on improving and optimizing an already known design. Re-engineering may involve improving component geometry, selecting better materials, modifying sealing arrangements, enhancing cooling or lubrication systems, and reducing maintenance requirements.

The goal is to enhance the performance of the existing system without completely redesigning it from the beginning.

7.3 Why Re-Engineering Becomes Necessary

In industrial environments, equipment is often required to operate for many years under varying conditions. Over time, several factors create the need for re-engineering:

Changing Operating Conditions

Industrial processes may evolve, leading to changes in operating pressure, temperature, fluid properties, or shaft speed. A seal design that was originally suitable may no longer perform effectively under new conditions. Process intensification, capacity expansions, and changes in product formulations are common triggers for re-engineering needs.

Equipment Upgrades

When machinery is upgraded to operate at higher capacities or speeds, the original sealing system may require modification to handle increased loads and thermal stresses. The seal must be re-evaluated as part of any significant equipment modification programme.

Frequent Seal Failures

Repeated seal failures indicate that the original design may not be well suited to the operating environment. Re-engineering can identify the root cause and implement improvements that permanently eliminate the recurring failure, rather than simply repeating the replacement cycle.

Improved Materials and Technologies

Advancements in materials and manufacturing technologies often create opportunities to improve existing designs. Modern ceramic and carbide materials may provide better wear resistance than materials used in older designs. Adopting these advances through re-engineering can significantly extend seal life.

Environmental and Safety Regulations

Stricter environmental regulations may require sealing systems that minimize leakage of hazardous fluids. Re-engineering may be necessary to achieve improved containment standards and comply with updated legal requirements.

7.4 Re-Engineering Process in Mechanical Seal Systems

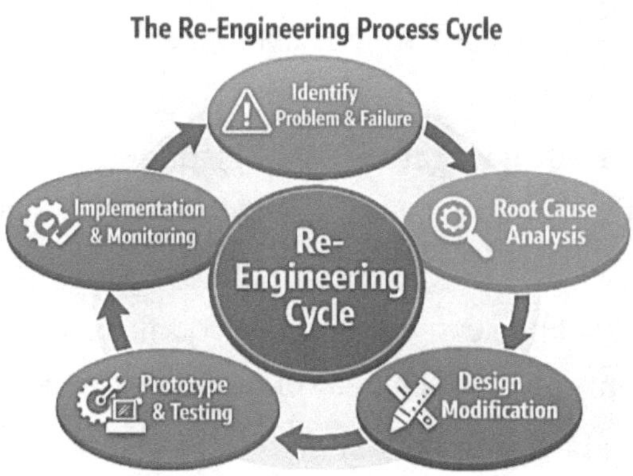

Re-engineering typically follows a structured engineering process:

1. **Performance Evaluation** — analysis of operational data, maintenance records, and failure history
2. **Failure Analysis** — examination of worn or damaged components to identify root cause
3. **Design Review** — evaluation of original design in light of current operating conditions
4. **Engineering Modifications — introduction of targeted changes to address identified shortcomings**
5. **Testing and Validation — confirmation that the improved design achieves the desired performance**

This structured approach ensures that re-engineering activities are driven by evidence and engineering understanding rather than guesswork. Each step builds on the previous one, creating a knowledge base that becomes a permanent asset for the organization.

7.5 Re-Engineering Opportunities in Mechanical Seal Applications

Several aspects of mechanical seals can be improved through re-engineering:

- **Seal face material improvement** — replacing older materials with advanced ceramics or carbides can improve wear resistance and thermal performance
- **Improved seal geometry** — adjusting face width, contact pressure distribution, or face alignment can enhance sealing stability
- **Enhanced cooling systems** — improved flush systems or cooling arrangements can reduce heat build-up and extend seal life
- **Improved elastomer selection** — selecting elastomers with better chemical compatibility can prevent swelling, hardening, or degradation

- **Cartridge seal conversion** — converting conventional component seals to cartridge seals can simplify installation and reduce alignment errors

7.6 Benefits of Re-Engineering

Re-engineering offers several benefits for industrial operations:

- **Improved equipment reliability** — design modifications reduce failure rates and increase operating stability
- **Extended seal life** — better materials and improved designs can significantly increase service life
- **Reduced maintenance costs** — longer seal life reduces the need for frequent maintenance and replacement
- **Improved process safety** — enhanced sealing performance minimizes leakage of hazardous fluids
- **Better energy efficiency** — optimized designs may reduce friction losses and improve equipment efficiency

7.7 Re-Engineering vs Complete Redesign

It is important to distinguish between re-engineering and complete redesign. Re-engineering focuses on improving existing designs through targeted modifications, while redesign involves developing an entirely new system.

Re-engineering is often preferred because it requires less development time, utilizes existing knowledge and equipment, and minimizes operational disruptions. However, if the original design has fundamental limitations that cannot be adequately addressed through modification, a complete redesign may sometimes be the more appropriate choice.

7.8 Role of Engineering Experience

Successful re-engineering requires both analytical skills and practical engineering experience. Engineers must be able to interpret operational data, recognize failure patterns, and understand how design changes will affect performance.

Field experience plays an important role in identifying practical solutions that may not be immediately evident through theoretical analysis alone. This is one of the reasons why experienced engineers — those who have encountered multiple failure scenarios across different applications — are particularly valuable in re-engineering activities.

7.9 Re-Engineering as a Step Toward Innovation

Re-engineering often leads to valuable insights that contribute to technological advancement. When engineers systematically analyse and improve existing designs, they may discover opportunities for new design concepts, improved materials, or advanced sealing technologies.

Thus, re-engineering serves as an important bridge between practical industrial experience and engineering innovation. Many of the most significant advances in mechanical seal technology have originated from systematic re-engineering activities that exposed fundamental limitations in previous designs.

7.10 Conclusion

Re-engineering is a powerful engineering methodology that allows engineers to improve existing systems and adapt them to changing industrial requirements.

By analyzing performance data, identifying failure causes, and implementing targeted design modifications, engineers can enhance the reliability and efficiency of mechanical seal systems. In many industrial applications, re-engineering provides a cost-effective solution for improving equipment performance without the need for complete system redesign.

The next chapter will explore specific re-engineering strategies applied to mechanical seals for performance improvement.

This page has been left blank intentionally.

Chapter 8

Re-Engineering Mechanical Seals for Performance Improvement

8.1 Introduction

Mechanical seals are critical components in rotating equipment such as pumps, compressors, mixers, and agitators. Although mechanical seals are designed to operate reliably under specified conditions, real industrial environments often present challenges that may affect seal performance.

These challenges may include changes in operating parameters, variations in fluid properties, equipment misalignment, or the presence of abrasive particles. In such situations, re-engineering becomes an effective engineering approach to improve seal performance without completely redesigning the system.

8.2 Identifying Performance Issues in Mechanical Seals

Before implementing re-engineering improvements, engineers must first identify the factors causing performance problems. Common performance issues in mechanical seals include excessive leakage, frequent seal failures, short seal life, overheating of seal faces, elastomer degradation, and excessive vibration or noise.

These problems may arise due to a combination of design limitations, operating conditions, or installation errors. A systematic engineering analysis is required to determine the root cause before introducing modifications.

8.3 Root Cause Analysis

Root cause analysis is an essential step in the re-engineering process. Engineers must investigate the underlying reasons behind seal failures rather than simply replacing damaged components. Typical root causes may include:

- **Improper material selection** — seal face materials or elastomers may not be compatible with the process fluid or temperature range
- **Inadequate lubrication** — insufficient fluid film between seal faces may cause excessive friction and wear, particularly in low-viscosity fluid applications
- **Thermal distortion** — excessive heat generation may cause seal faces to distort and lose proper contact, particularly in high-speed applications
- **Misalignment of equipment** — shaft misalignment or excessive runout may cause uneven loading on the seal faces and premature wear
- **Improper installation** — incorrect installation practices may damage seal components or cause uneven face contact from the first moment of operation

By identifying the root cause of the problem, engineers can develop effective re-engineering solutions that provide permanent improvement rather than temporary relief.

8.4 Improving Seal Face Materials

One of the most common re-engineering strategies is improving the seal face material combination. Older seal designs may use materials that are no longer optimal for modern industrial conditions.

Replacing these materials with advanced alternatives can significantly improve seal performance. Examples include replacing ceramic faces with silicon carbide for improved wear resistance, using tungsten carbide for abrasive applications, and using improved grades of carbon graphite for better lubrication properties.

Material upgrades can improve resistance to wear, thermal distortion, and chemical attack simultaneously.

8.5 Optimizing Seal Face Geometry

Seal face geometry plays an important role in maintaining proper contact and lubrication conditions. Re-engineering may involve adjusting seal face width, improving surface finish and flatness, optimizing contact pressure distribution, and improving face alignment.

Proper geometry helps maintain a stable lubrication film and reduces frictional heat generation. In some applications, the introduction of hydrodynamic grooves on the seal face can provide additional separation and reduce wear significantly.

8.6 Improving Cooling and Lubrication

Excessive heat generation is one of the major causes of mechanical seal failure. Re-engineering the cooling or lubrication system can greatly improve seal performance. Possible improvements include:

- **Improved flush systems** — introducing a controlled flow of clean fluid near the seal faces to remove heat and maintain lubrication, following recognised API piping plans

- **Seal chamber modifications** — changing the seal chamber geometry to improve fluid circulation around the seal and prevent vapour pocket formation
- **External cooling arrangements** — in high-temperature applications, external cooling systems such as heat exchangers or water jackets may be used to control seal temperature

8.7 Improving Secondary Sealing Elements

Secondary sealing elements such as O-rings and gaskets are critical for preventing leakage around seal components. Re-engineering may involve selecting elastomers that offer better resistance to high temperatures, chemical attack, and mechanical deformation.

For example, replacing nitrile rubber with fluoroelastomers or PTFE-based materials may improve performance in aggressive chemical environments. The chemical compatibility of secondary seals is as important as that of the primary seal faces.

8.8 Improving Installation and Alignment

In many cases, mechanical seal failures occur not because of design problems but due to improper installation or equipment misalignment.

Re-engineering solutions may include introducing cartridge mechanical seals for easier and more reliable installation, improving shaft alignment procedures, reducing shaft runout and vibration through bearing maintenance, and providing better training for maintenance personnel.

Cartridge seals eliminate most installation errors by pre-setting the correct spring compression at the factory, removing a major source of field-based failures.

8.9 Adapting Seals to Changing Process Conditions

Industrial processes often change over time. Equipment may be required to operate at higher capacities, different temperatures, or with different fluids. Re-engineering allows engineers to adapt existing seal designs to these new conditions.

Possible modifications include upgrading seal materials to handle higher temperatures, modifying spring arrangements to accommodate higher pressures, and changing seal configurations for improved containment.

Through such modifications, existing equipment can continue to operate reliably under new conditions without the cost of complete replacement.

8.10 Reducing Maintenance Costs

One of the major objectives of re-engineering is to reduce maintenance costs associated with mechanical seal failures. By improving seal design and reliability, plants can achieve longer seal life, reduced frequency of seal replacement, lower downtime for equipment maintenance, and reduced spare parts inventory.

The cumulative cost savings from a successful re-engineering programme — even on a single critical pump — can be substantial over the lifetime of the equipment.

8.11 Role of Re-Engineering in Industrial Reliability

Re-engineering plays a vital role in improving the reliability of industrial equipment. Rather than accepting recurring failures as unavoidable, engineers can analyse the underlying causes and implement design improvements. This approach transforms maintenance

activities from reactive problem-solving into proactive reliability engineering. In many cases, relatively small modifications in seal design or operating conditions can produce significant improvements in equipment performance.

8.12 Re-Engineering as a Knowledge-Building Process

Every re-engineering activity generates valuable engineering knowledge. By studying equipment performance and implementing improvements, engineers develop deeper understanding of the interaction between design, materials, and operating conditions.

This accumulated knowledge becomes a valuable resource for future engineering work and contributes to technological advancement. Re-engineering therefore forms an important link between practical industrial experience and engineering innovation.

8.13 Conclusion

Re-engineering provides an effective and practical method for improving the performance and reliability of mechanical seal systems used in industrial equipment. By analyzing existing designs, identifying root causes of failures, and implementing targeted improvements, engineers can enhance seal life, reduce maintenance costs, and improve operational safety.

Re-engineering also contributes to the continuous evolution of sealing technology by transforming operational experience into engineering knowledge. In the next chapter, we will explore the concept and methodology of reverse engineering.

PART IV

Reverse Engineering of Mechanical Seals

Chapters 9 — 11

This page has been left blank intentionally.

Chapter 9

Understanding Reverse Engineering

9.1 Introduction

In modern engineering practice, industries frequently encounter situations where the design details of a component or system are not fully available. This may occur when equipment documentation is incomplete, when spare parts are no longer supported by the original manufacturer, or when engineers need to study an existing product in order to improve its performance. In such situations, engineers often apply a systematic engineering approach known as Reverse Engineering.

Reverse engineering involves analyzing an existing product, component, or system in order to understand its design, materials, manufacturing methods, and functional principles.

The objective of reverse engineering is not merely to replicate a product, but to gain technical understanding of how the product works and how it can be improved or adapted for specific applications.

In mechanical seal technology, reverse engineering can play an important role in studying existing seal designs, analyzing failures, and developing improved sealing solutions.

9.2 Definition of Reverse Engineering

Reverse Engineering can be defined as the systematic process of examining and analyzing an existing product to understand its design structure, materials, dimensions, and functional principles.

Unlike Ready Engineering, which starts with engineering principles to develop new designs, reverse engineering begins with an existing product and works backward to understand how the design was created.

The process typically involves detailed measurement of components, material identification, study of manufacturing processes, functional analysis of the system, and evaluation of operating conditions.

9.3 Reverse Engineering vs Simple Copying — The Critical Distinction

One of the most common misunderstandings about reverse engineering is the belief that it simply involves copying an existing product. This perception is incorrect and does a serious disservice to the engineering discipline.

True reverse engineering is far more complex and requires substantial engineering expertise.

Simple copying focuses only on replicating external dimensions or visible features of a product. This approach ignores the deeper engineering considerations that influence performance and reliability.

A copied seal may look identical to the original, yet fail within days or weeks of installation, because the copier did not understand the operating philosophy behind the original design.

Reverse engineering, by contrast, involves understanding the complete engineering logic behind the design, including:

- **Why specific materials were selected** — what operating conditions drove that choice
- **How mechanical loads are distributed** — what the stress state of each component is
- **How thermal conditions affect performance** — what the thermal management strategy is
- **How fluid interactions influence the design** — what the hydrodynamic and pressure balance principles are

Without understanding these factors, a copied product may fail even if its dimensions appear identical to the original. The failure mode will often be different from the original failure being investigated, making diagnosis even more difficult.

Copying dimensions ≠ Reverse Engineering
Understanding the engineering logic = Reverse Engineering

9.4 Engineering Effort in Reverse Engineering

Successful reverse engineering requires a high level of engineering capability. Engineers must combine analytical skills, practical experience, and technical knowledge to accurately interpret the design of an existing product. The engineering effort involved includes:

- **Dimensional Analysis** — precise measurement of components using advanced instruments such as coordinate measuring machines (CMM), micrometers, and optical scanners
- **Material Identification** — determining the material composition of components using metallurgical analysis, hardness testing, or chemical analysis
- **Functional Analysis** — studying how the component functions within the overall system and how it interacts with other components
- **Performance Evaluation** — analyzing how the design performs under operating conditions such as pressure, temperature, and rotational speed
- **Manufacturing Analysis** — understanding how the component was manufactured, including machining processes, surface finishing, and heat treatment

These activities require both technical expertise and careful engineering judgment. The effort invested in a thorough reverse engineering exercise is considerable — but it is this effort that separates true reverse engineering from mere copying.

9.5 Role of Reverse Engineering in Mechanical Seal Technology

Reverse engineering has several practical and legitimate applications in mechanical seal engineering:

- **Analysis of existing seal designs** — engineers may study commercially available seal designs to understand their construction and performance characteristics, informing future design work

- **Replacement of obsolete components** — when original equipment manufacturers no longer supply spare parts, reverse engineering allows engineers to reproduce these components for operational continuity
- **Failure investigation** — by analyzing damaged seal components, engineers can identify the causes of failure and develop improved designs
- Adaptation for new applications — existing seal designs may be modified through reverse engineering to suit different equipment or operating conditions

9.6 Ethical Considerations in Reverse Engineering

Reverse engineering must be practiced responsibly and ethically. Engineers must respect intellectual property rights, patents, and legal restrictions when analysing and reproducing existing designs. Ethical reverse engineering focuses on technical understanding, improvement, and innovation, rather than unauthorized duplication of proprietary products. Responsible engineering practices include:

- Respecting intellectual property laws and patent protections
- Avoiding infringement of patented designs — understanding a design does not confer the right to reproduce it commercially if it is protected
- Using reverse engineering primarily for analysis, learning, and improvement
- Ensuring that any reproduced components meet the safety and performance requirements of the application

Professional engineers have a responsibility to ensure that their work adheres to both legal and ethical standards. The engineering profession derives its credibility from this integrity.

9.7 Importance of Engineering Knowledge in Reverse Engineering

Reverse engineering requires a strong foundation in engineering fundamentals. Without sufficient technical knowledge, engineers may misinterpret the purpose of certain design features or fail to recognize critical aspects of the system.

Engineering knowledge helps interpret stress distribution in components, thermal behavior of materials, fluid flow characteristics, and tribological interactions between surfaces. In mechanical seal systems, even small design features may have important functional purposes that are not immediately obvious.

Understanding these features requires careful engineering analysis.

9.8 Reverse Engineering as a Learning Process

Reverse engineering can be a powerful educational tool for engineers. By studying existing designs, engineers gain valuable insights into practical engineering solutions and design strategies.

This process allows engineers to understand successful design principles, learn from past engineering experience, identify design limitations, and develop improved solutions. Many technological innovations have emerged from engineers studying existing systems and discovering ways to improve them.

9.9 Reverse Engineering and Industrial Innovation

Reverse engineering can contribute significantly to technological advancement when used as a tool for learning and improvement. By analyzing existing products, engineers may discover opportunities for improved materials, better manufacturing methods, enhanced performance, and greater reliability.

When these insights are applied creatively, they lead to innovation in engineering design. Thus, reverse engineering should not be viewed as a backward-looking activity, but as an important step in the continuous development of engineering knowledge.

9.10 Reverse Engineering in the Age of Artificial Intelligence

Modern technologies such as 3D scanning, computer-aided design, and artificial intelligence have greatly enhanced the capabilities of reverse engineering. Advanced digital tools allow engineers to create precise digital models of components, analyse complex geometries, simulate operating conditions, and evaluate design performance.

Artificial intelligence can assist engineers in analyzing large datasets and identifying patterns in design performance.

However, even with these advanced tools, human engineering judgment remains essential for interpreting results and making design decisions. AI can tell an engineer what a surface looks like in three dimensions; it cannot tell the engineer why that geometry was chosen or whether a different geometry would perform better under the specific operating conditions of the application. ***That judgment belongs to the engineer.***

9.11 Conclusion

Reverse engineering is a powerful engineering methodology used to analyse and understand existing products and systems. By examining design features, materials, and functional behavior, engineers can gain valuable technical knowledge and develop improved solutions.

Unlike simple copying, true reverse engineering requires deep engineering understanding and analytical capability. When practiced responsibly and ethically, reverse engineering becomes an important tool for learning, improvement, and innovation in engineering technology. The next chapter will explore the detailed methodology of reverse engineering as applied to mechanical seals.

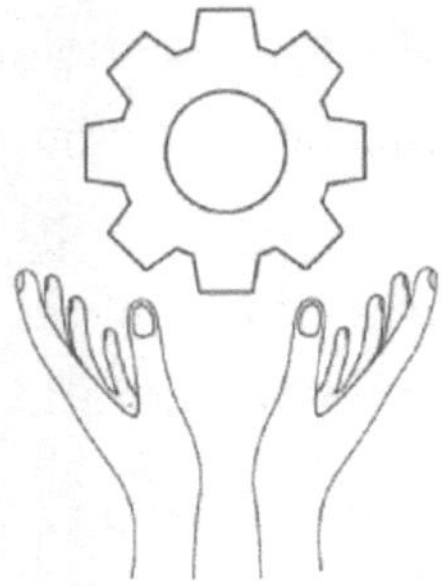

Chapter 10

Reverse Engineering Methodology for Mechanical Seals

10.1 Introduction

Reverse engineering of mechanical seals involves a systematic and disciplined engineering process aimed at understanding the design, materials, geometry, and functional characteristics of an existing seal system.

Mechanical seals are precision components used in rotating equipment, and their successful operation depends on the interaction of multiple elements such as seal faces, springs, elastomers, and metal components. Therefore, reverse engineering of such systems requires careful analysis and accurate interpretation of both physical and functional features.

The objective of reverse engineering in mechanical seal technology is not merely to replicate a seal, but to understand the engineering principles behind its design, evaluate its performance, and potentially develop improved solutions.

10.2 Overview of the Reverse Engineering Process

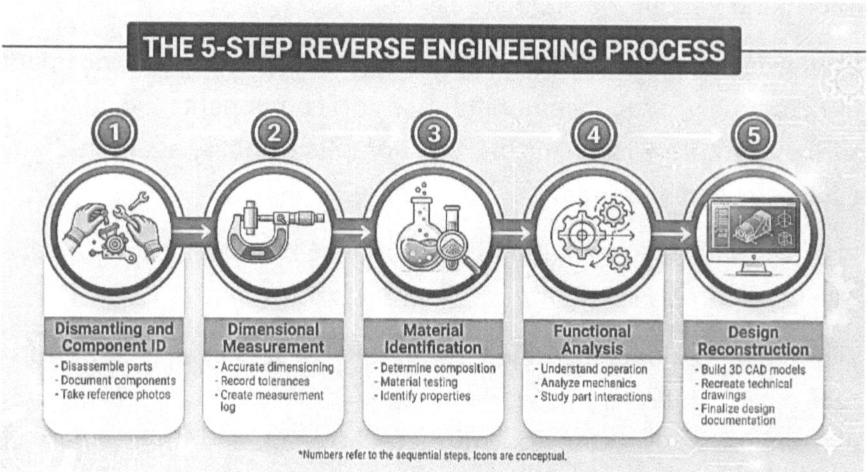

The reverse engineering process for mechanical seals generally involves five major steps:

1. Dismantling and component identification
2. Detailed dimensional measurement
3. Material identification and characterization
4. Functional analysis of components
5. Engineering reconstruction of the design

Each step requires careful documentation and systematic engineering evaluation. Skipping any step — or treating it as a formality — risks creating a flawed reconstruction that will fail in service.

10.3 Step 1 — Dismantling and Component Analysis

The first stage in reverse engineering is the careful dismantling of the mechanical seal assembly. Mechanical seals typically consist of multiple components assembled together with precise alignment and specific tolerances.

Improper dismantling may damage parts or lead to loss of important information about the original design.

During dismantling, engineers should carefully observe and document the assembly sequence, orientation of components, location of springs and drive mechanisms, position of secondary seals, and any evidence of wear or damage.

Photographic documentation is essential at this stage to record the original assembly configuration before it is disturbed.

Typical components identified during this stage include the rotating seal face, stationary seal face, springs or bellows, O-rings or elastomeric seals, retainers and metal housing, and drive collars or locking mechanisms. Each component should be labelled, measured, and stored carefully for subsequent analysis.

10.4 Step 2 — Dimensional Measurement Techniques

Once the components are separated and identified, the next step is to determine the exact dimensions and geometry of each component. Accurate dimensional measurements are essential for understanding the design tolerances and reconstructing the original engineering drawings.

Common measurement instruments used include:

- Vernier calipers — for overall dimensions and basic geometry
- Micrometers — for precise diameter and thickness measurements
- Depth gauges — for shoulder depths and bore measurements
- Dial indicators — for runout and alignment measurements
- Surface roughness testers (profilometers) — for characterising seal face finish
- Coordinate Measuring Machines (CMM) — for complex geometries and tight tolerance verification

- Optical scanners and 3D laser scanners — for creating complete digital representations of components

Important dimensions that must be measured include seal face diameter and face width, component thickness and overall length, spring dimensions and free length, shaft interface dimensions, housing fit dimensions, and the critical flatness of seal faces. All measurements should be carefully recorded and verified to ensure accuracy.

10.5 Step 3 — Material Identification Methods

Understanding the materials used in the seal components is a critical step in reverse engineering. Identifying the materials helps engineers understand the design choices made in the original seal and enables the selection of appropriate replacement materials.

Material identification may involve several techniques:

- **Visual and physical inspection** — initial identification based on appearance, density, and surface characteristics (dark matte for carbon, high density for carbide, white for ceramic)
- **Hardness testing** — Vickers or Rockwell hardness measurements provide useful information about material properties
- **Metallurgical analysis** — optical microscopy and scanning electron microscopy (SEM) reveal microstructural features and alloy characteristics
- **X-ray fluorescence (XRF) spectroscopy** — non-destructive chemical composition analysis for metals and some ceramics
- **Elastomer identification** — FTIR spectroscopy, immersion tests, or hardness measurements to distinguish between NBR, FKM, EPDM, and PTFE-based materials

10.6 Step 4 — Functional Analysis of Seal Components

After the physical characteristics of the components have been determined, the next step is to understand how each component contributes to the overall sealing function. Functional analysis involves examining the interaction between components and evaluating how the design maintains effective sealing under operating conditions.

Important aspects of functional analysis include:

- **Seal Face Interaction** — studying the contact conditions between the rotating and stationary seal faces to understand how friction and lubrication are controlled
- **Spring Loading Mechanism** — analyzing the spring arrangement, stiffness, and preload conditions to understand the closing force philosophy
- **Secondary Sealing** — examining the role of elastomeric elements in preventing leakage between seal components and equipment surfaces
- **Drive Mechanism** — understanding how the drive pins, set screws, or collars transmit rotational motion from the shaft to the rotating seal face
- **Pressure Balance** — evaluating the hydraulic balance ratio to understand how the design manages the hydraulic closing and opening forces
- **Heat and Fluid Interaction** — evaluating how the design allows fluid to circulate near the seal faces to provide lubrication and cooling

10.7 Step 5 — Engineering Reconstruction of the Design

After dimensional measurements, material identification, and functional analysis have been completed, engineers can proceed to reconstruct the design of the mechanical seal. This stage involves creating engineering drawings or digital models that represent the original seal configuration.

Computer-aided design software is used to generate accurate models of each component and the complete assembly.

The reconstruction process includes creating detailed component drawings with all critical dimensions and tolerances, defining material specifications for each component, reconstructing the assembly structure with correct fits and clearances, and documenting the design intent as understood through functional analysis.

This reconstructed design provides a complete technical representation of the seal system that can serve as the basis for reproduction, improvement, or adaptation.

10.8 Evaluation of Design Performance

Once the design has been reconstructed, engineers may evaluate its expected performance under operating conditions. This evaluation may include analysis of contact pressure between seal faces, estimation of frictional heat generation, evaluation of material compatibility with process fluids, and assessment of mechanical stresses in components.

Through this evaluation, engineers may identify opportunities for improving the design beyond the original specification.

10.9 Opportunities for Design Improvement

Reverse engineering often reveals opportunities to enhance the original design. Possible improvements may include upgrading seal face materials, improving cooling or lubrication arrangements, optimizing spring loading conditions, and enhancing resistance to corrosion or wear.

Such improvements may lead to re-engineered seal designs with better reliability and performance than the original. This is the highest outcome of a successful reverse engineering exercise — the original product is not merely reproduced, but improved.

10.10 Importance of Documentation

Accurate documentation is essential throughout the reverse engineering process. Engineers should maintain detailed records of component photographs, measurement data, material analysis results, functional observations, and reconstructed design drawings.

This documentation ensures that the engineering knowledge gained through reverse engineering is preserved and can be used for future reference, further improvement, or training of younger engineers.

10.11 Conclusion

Reverse engineering of mechanical seals is a systematic engineering process that involves careful dismantling, precise measurement, material identification, functional analysis, and design reconstruction.

Through this methodology, engineers can gain a comprehensive understanding of existing seal designs and evaluate their performance characteristics. The insights gained from reverse engineering can contribute to improved seal designs, better maintenance practices, and enhanced reliability of industrial equipment.

10.12 Ethical Considerations in Reverse Engineering

The methodology described in this chapter is powerful, and like all powerful engineering tools, it must be applied responsibly. This section addresses the ethical framework within which reverse engineering must be practiced.

10.12.1 Reverse Engineering Is Not Copying

True reverse engineering is an engineering learning process that requires deep analysis, interpretation, and technical understanding. A responsible engineer performing reverse engineering must understand the functional purpose of each component, analyse material selection rationale, evaluate manufacturing feasibility, study the operating conditions, and identify limitations of the original design.

Only after this engineering analysis can a new design be reconstructed. In most cases, the final design developed through reverse engineering differs significantly from the original product, because the engineer introduces improvements based on better understanding and experience.

10.12.2 Respect for Engineering Effort

Every engineered product represents significant effort in terms of research, experimentation, design calculations, material development, and manufacturing processes. Even when reverse engineering an existing product, engineers must recognise and respect the technical effort that went into the original design. The objective should not be to imitate, but to learn, improve, and innovate.

10.12.3 Reverse Engineering for Engineering Advancement

Ethical reverse engineering contributes positively to technological development. It helps industries understand complex engineering systems, develop alternative solutions, improve reliability, adapt designs for different operating environments, and reduce excessive dependency on proprietary suppliers.

In many cases, reverse engineering enables local industries to develop indigenous engineering capabilities, which has significant national and economic importance.

10.12.4 Professional Responsibility of Engineers

Engineers involved in reverse engineering must follow certain professional principles: maintaining technical honesty in analysis and reporting; avoiding direct copying of proprietary designs without engineering modification; ensuring that reconstructed designs meet safety

and performance requirements; and using reverse engineering as a learning and development tool rather than a shortcut.

Engineering ethics demand that reverse engineering should always lead to knowledge development and technological improvement.

10.12.5 Final Perspective on Ethics

When practiced with proper engineering knowledge and ethical responsibility, reverse engineering becomes a valuable tool for technological progress and industrial self-reliance. The engineer who approaches reverse engineering with intellectual honesty and professional integrity elevates it from a technical exercise to a contribution to the advancement of the engineering profession itself.

Reverse Engineering + Engineering Intelligence + Ethics = Innovation

Chapter 11

Failure Analysis as a Tool for Reverse Engineering

11.1 Introduction

Mechanical seals are precision-engineered components designed to operate under demanding conditions involving pressure, temperature, speed, and chemical exposure.

Despite careful design and manufacturing, seal failures can still occur due to various reasons related to equipment condition, operating environment, installation practices, or incorrect seal selection. Failure analysis is therefore an essential part of mechanical seal engineering.

The objective of failure analysis is not merely to identify that a seal has failed, but to determine why the failure occurred and how similar failures can be prevented in the future.

A systematic failure analysis helps engineers improve seal design, improve equipment reliability, reduce maintenance costs, increase operational safety, and extend seal life. Failure analysis is therefore a valuable learning process that contributes directly to engineering improvement and innovation.

It is also one of the most powerful inputs to a reverse engineering exercise, because it reveals not only what the original design contained, but where it was inadequate.

11.2 Importance of Systematic Failure Analysis

Many organizations treat seal failure as a routine maintenance event and simply replace the failed seal with a new one. However, this approach does not address the root cause of failure, and without proper investigation, the same failure is likely to repeat, maintenance costs continue to increase, and equipment reliability decreases.

A systematic failure analysis involves collection of operating data, inspection of failed seal components, understanding of equipment condition, study of process parameters, and engineering evaluation of the failure mechanism. Only after understanding the true root cause can a permanent solution be implemented.

11.3 Common Causes of Mechanical Seal Failure

Mechanical seal failures generally occur due to one or more of the following factors.

Improper Installation

Incorrect installation is one of the most common causes of early seal failure. Typical installation-related issues include incorrect setting length, improper tightening of fasteners, misalignment during installation, damage to seal faces during handling, and contamination entering the seal chamber.

Even a well-designed seal can fail quickly if it is not installed correctly. This is one of the strongest arguments in favor of cartridge seal designs, which eliminate most installation variables.

Equipment Misalignment

Misalignment between rotating and stationary components creates excessive stress on the seal faces. Sources of misalignment include shaft deflection, bearing wear, improper coupling alignment, and thermal distortion of equipment.

Misalignment can lead to uneven seal face wear, excessive heat generation, and leakage. In many industrial plants, pumps are realigned after installation and commissioning but are not checked again after years of operation and bearing replacements. This is a common oversight.

Dry Running

Mechanical seals depend on a thin lubricating film between seal faces. When this film disappears, direct contact occurs between the seal faces, and catastrophic failure typically follows within minutes or seconds.

Dry running may occur due to lack of fluid in the seal chamber, air pockets in the system, pump running without liquid, or improper flushing arrangements. The damage caused by dry running is often unmistakable in its visual characteristics.

Excessive Heat Generation

Heat generation at the seal faces can lead to thermal distortion, material degradation, and failure of secondary sealing elements. Heat may increase due to inadequate cooling, high

friction, improper face materials, high rotational speed, or poor lubrication. Thermal effects often result in seal face cracking or deformation — sometimes described as heat checking.

Abrasive Wear

In applications involving slurry or contaminated fluids, abrasive particles may enter the seal interface. Abrasive particles can cause rapid wear of seal faces, scoring of surfaces, loss of flatness, and leakage. Slurry pumping applications are particularly challenging and require special seal designs, material selections (typically hard face on hard face), and flushing arrangements.

Chemical Attack

In chemical processing industries, the process fluid may react with seal materials, affecting elastomers, seal face materials, or metal components. Material compatibility is a critical consideration during seal selection, and this type of failure often results from a change in the process chemistry that was not communicated to the seal engineering team.

Pressure Fluctuations

Sudden pressure changes may cause instability in seal operation. Pressure fluctuations may occur due to pump cavitation, system pressure surges, or improper pressure balancing. These conditions may lead to seal face separation or instability, resulting in rapid wear and leakage.

11.4 Visual Inspection of Failed Seals

Visual examination of the failed seal components often provides valuable clues regarding the failure mechanism. Typical inspection areas include seal face condition, wear patterns, cracking, discoloration, scoring marks, and elastomer degradation. Each of these signs provides information about the operating conditions during failure:

- **Bluish or rainbow discoloration** — indicates overheating at the seal face interface
- **Radial scoring or grooving** — indicates abrasive contamination entering the seal interface

- **Uneven wear across the face width** — indicates misalignment or shaft runout
- **Blistering or pitting on carbon face** — indicates dry running or vapour flashing
- **Hardened or cracked elastomers** — indicates chemical incompatibility or excessive temperature
- **Corroded metal components** — indicates chemical attack from the process fluid or environment

Careful observation helps engineers reconstruct the failure sequence and identify the primary initiating cause.

11.5 Seal Face Failure Patterns

Seal face damage often follows recognizable patterns that are diagnostic of specific failure causes:

- **Uniform narrow wear track** — typically indicates normal operation with some misalignment or thermal distortion reducing the effective contact width
- **Localized wear at one angular position** — indicates shaft runout or dynamic misalignment
- **Thermal cracking** (heat checking) radial cracks from rapid thermal cycling, typically caused by loss of lubrication or intermittent dry running
- **Blistering** — trapped gases or excessive heat causing surface disruption, common in high-vapor-pressure applications
- **Chemical etching or corrosion** — attack by process fluid, particularly visible on carbon or ceramic faces

11.6 Importance of Operating Data in Failure Analysis

Failure analysis should always include operating data such as operating pressure, temperature, rotational speed, fluid characteristics, equipment vibration levels, and seal chamber design.

Without this information, it becomes difficult to understand the actual conditions under which the seal was operating. Field data combined with physical inspection provides the most reliable basis for failure diagnosis.

Engineers should always request this information before rendering a conclusion, and organizations should maintain proper operating records for their sealing equipment.

11.7 Engineering Approach to Failure Prevention

Once the root cause of failure has been identified, engineers can implement corrective measures. These may include modification of seal design, selection of improved materials, improved installation procedures, equipment alignment correction, improved flushing systems, and improved cooling arrangements.

Preventive engineering solutions are always more effective than repeated maintenance replacements. A seal that is re-engineered based on systematic failure analysis will deliver far better service than one that is simply replaced with an identical unit.

11.8 The Mechanical Seal Failure Pyramid

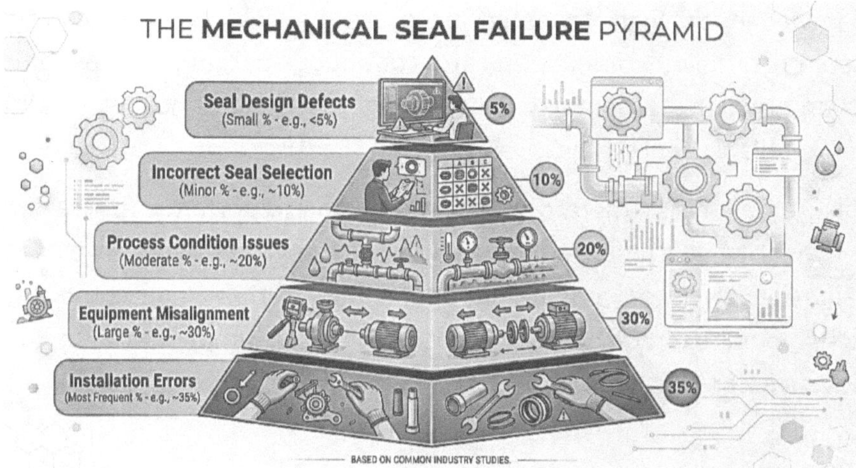

Industry experience accumulated over decades consistently demonstrates that the majority of mechanical seal failures are not caused by the seal design itself, but by the broader system and application factors. This insight is captured in the Mechanical Seal Failure Pyramid:

The pyramid shows that installation and handling practices — factors entirely within the control of the operating organization — account for the largest share of failures. The lesson is clear: investing in proper installation training and procedures, maintaining equipment alignment, and thoroughly understanding the application before selecting a seal will prevent far more failures than switching to a more expensive seal design.

The engineering lesson from this pyramid is also fundamental to reverse engineering: when analyzing a failed seal, the engineer must look beyond the seal itself and examine the entire system context. The seal is rarely the sole cause of its own failure.

11.9 Learning from Failures

Every failure provides an opportunity to improve engineering understanding. Experienced seal engineers often develop their knowledge through repeated exposure to field failures, careful analysis of failed components, and continuous design improvement. Over time, this accumulated knowledge contributes to better and more reliable seal designs.

Failure analysis therefore becomes an important source of engineering learning and innovation. An engineer who embraces failure analysis as a learning opportunity rather than a blame-assignment exercise will grow faster and contribute more to the advancement of the field.

11.10 Final Engineering Perspective

Mechanical seal failures should never be viewed simply as maintenance problems. They are engineering feedback mechanisms that help improve design, application, and operational practices.

Through systematic failure analysis, engineers can transform failures into valuable learning opportunities that lead to improved designs, increased equipment reliability, and enhanced operational efficiency. In this way, failure analysis becomes an integral part of engineering intelligence and continuous technological improvement.

The next chapter will examine the advanced tools and technologies used in modern reverse engineering practice.

This page has been left blank intentionally.

PART V

Modern Tools and Technologies

Chapters 12 — 13

This page has been left blank intentionally.

Chapter 12

Advanced Reverse Engineering Tools and Technologies

12.1 Introduction

The practice of reverse engineering has been significantly enhanced in recent decades by the development of advanced measurement, analysis, and modelling technologies. These tools allow engineers to characterize existing components with greater accuracy, speed, and completeness than was previously possible.

For mechanical seal reverse engineering, where dimensional tolerances are measured in micrometers and surface finishes in nanometers, these technologies are particularly valuable.

However, it is important to emphasize that tools alone do not constitute reverse engineering. They are instruments in the hands of the engineer. Without the engineering knowledge to interpret the data these tools produce, they merely generate numbers without meaning.

The following sections describe the principal tools available to the modern reverse engineering engineer.

12.2 3D Scanning and Optical Measurement

Three-dimensional scanning technologies have transformed the speed and completeness with which physical objects can be digitally captured. White-light structured light scanners, photogrammetry systems, and laser scanners can create detailed three-dimensional point clouds of seal components within minutes.

These digital representations capture the complete geometry of a component, including features that would be difficult or time-consuming to measure individually with conventional instruments.

For mechanical seals, 3D scanning is particularly valuable for capturing complex geometries such as drive slots, spring retainer profiles, and gland features. The digital model produced can be directly imported into CAD software for analysis and drawing generation.

However, the precision of 3D scanning is typically in the range of 0.01 to 0.05 mm, which may not be sufficient for the most critical dimensions such as seal face flatness. Conventional precision instruments remain essential for these critical measurements.

12.3 Coordinate Measuring Machines (CMM)

Coordinate Measuring Machines represent the highest level of dimensional measurement precision available for reverse engineering work. A CMM uses a probing system to measure the exact coordinates of points on a component surface in three-dimensional space, typically with an accuracy of a few micrometers. For mechanical seal components, CMM measurement can capture:

- Shaft bore diameter and tolerance with high precision
- Seal face diameter and face width with sub-micrometer accuracy
- Perpendicularity of seal face relative to shaft bore — critical for proper face contact
- Spring seat and drive slot dimensions

- Housing and gland features

CMM measurement data can be compared directly with reconstructed CAD models to verify dimensional accuracy and ensure that the reverse-engineered design meets the required tolerances.

12.4 CAD Modelling for Reconstruction

Computer-Aided Design software is the central tool in the reconstruction phase of reverse engineering. Once dimensional measurements and material data have been gathered, the engineer uses CAD software to create precise three-dimensional models of each component and the complete seal assembly. Modern CAD systems allow engineers to:

- Create parametric models that can be easily modified for design improvements
- Perform interference checks to verify correct assembly clearances
- Generate complete engineering drawings with dimensions, tolerances, and material specifications.
- Conduct basic finite element analysis to evaluate stress distribution in components
- Create assembly instructions and installation documentation

The quality of the CAD reconstruction depends critically on the quality of the measurement data and the engineering judgment applied in interpreting that data.

12.5 Material Characterization Techniques

Accurate material identification is essential for a meaningful reverse engineering exercise. The following laboratory techniques are used for mechanical seal component characterization:

- **X-ray Fluorescence (XRF) Spectroscopy** — provides non-destructive elemental composition analysis for metallic components; suitable for identifying stainless steel grades, carbide binder compositions, and alloy identifications
- **Scanning Electron Microscopy (SEM) with Energy Dispersive Spectroscopy (EDS)** — provides microstructural imaging and localized elemental analysis; particularly useful for ceramic and carbide face materials
- **Vickers Hardness Testing** — standard method for characterizing seal face hardness, which helps distinguish between different grades of carbide and ceramic materials
- **Fourier Transform Infrared Spectroscopy (FTIR)** — standard method for polymer and elastomer identification; can distinguish between NBR, FKM, EPDM, PTFE, and other elastomers definitively
- **Density Measurement** — a simple but often informative technique; different seal face materials have characteristic densities that aid identification

12.6 Quality Verification Methods

Once a component has been re-engineered and manufactured, it must be verified against the reconstructed design before it enters service. Quality verification for mechanical seal components typically involves:

- Dimensional inspection using CMM or precision hand instruments
- Surface flatness verification using optical flat and monochromatic light — for seal faces, flatness better than 1 to 3 helium light bands (0.3 to 0.9 micrometres) is typically required
- Surface roughness measurement using contact profilometry — seal faces typically require Ra values of 0.1 to 0.4 micrometres depending on the application

- Hardness testing of face materials to confirm specification compliance
- Visual inspection for cracks, chips, or surface defects

These verification steps are essential for ensuring that the reverse-engineered component will perform reliably in service. A component that appears correct dimensionally but has surface finish or flatness outside specification will not seal properly.

12.7 Digital Twin Concept in Seal Engineering

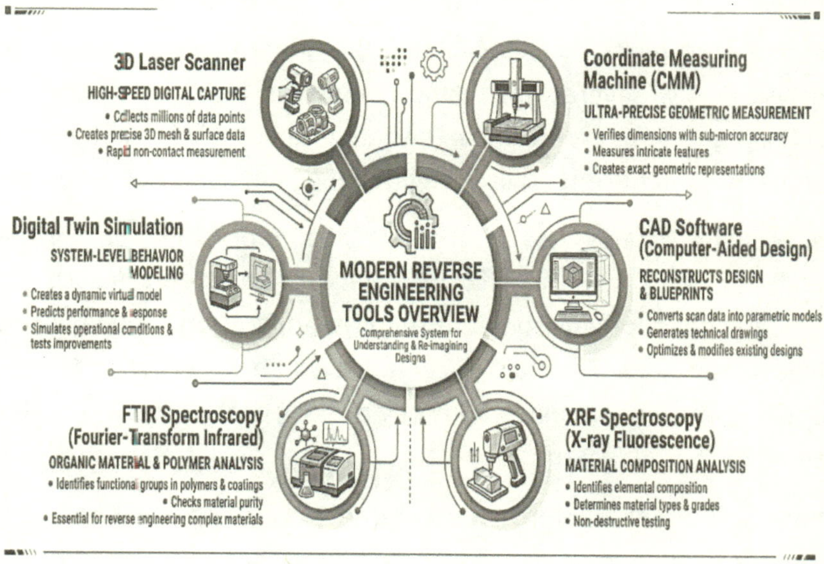

A digital twin is a virtual representation of a physical component or system that is continuously updated with real operating data from the physical counterpart. In mechanical seal engineering, the digital twin concept offers significant potential for performance monitoring and predictive maintenance.

A digital twin of a sealing system would integrate the geometric model of the seal, the operating conditions (pressure, temperature, speed), the physical properties of the materials, and the real-time sensor data from the installed seal.

This integrated model can be used to monitor seal health in real time, predict remaining useful life based on current operating conditions, identify developing problems before they cause failure, and optimize operating conditions to extend seal life.

While the full implementation of seal digital twins is still evolving, the underlying engineering — dimensional accuracy, material characterization, and functional analysis — is exactly the work done in a thorough reverse engineering exercise.

12.8 Limitations of Technology in Reverse Engineering

Despite the impressive capabilities of modern measurement and analysis tools, several important limitations must be recognised. Technology cannot determine design intent — why a particular geometry was chosen requires engineering judgment, not a scanner.

Technology cannot assess the original operating context — understanding that a specific material was selected because of a particular chemical environment requires application knowledge, not dimensional data. Technology cannot evaluate whether a design is good or merely functional — only engineering experience and analysis can make that judgment.

These limitations reinforce the central message of this book: tools enhance engineering capability, but they cannot substitute for engineering intelligence. The most sophisticated reverse engineering laboratory in the world, staffed by technicians without engineering understanding, will produce inadequate results. The same laboratory, used by experienced engineers, becomes a powerful instrument for innovation.

12.9 Conclusion

Advanced reverse engineering tools have significantly enhanced the engineer's ability to characterize, reconstruct, and improve existing mechanical seal designs. From 3D scanning and CMM measurement to material spectroscopy and digital twin modelling, these technologies provide powerful support for the engineering process.

However, they are most effective when used by engineers who possess the fundamental knowledge to interpret their outputs and exercise sound engineering judgment in applying those insights. The next chapter will examine the role of Artificial Intelligence in mechanical seal engineering.

This page has been left blank intentionally.

Chapter 13

Artificial Intelligence in Mechanical Seal Engineering

13.1 Introduction

Artificial Intelligence is increasingly being discussed across all engineering domains. Many industries are exploring how AI can assist in design, analysis, prediction, and optimization.

However, in highly specialized engineering fields such as mechanical sealing, AI should be viewed as a supporting tool rather than a replacement for engineering expertise.

Mechanical seals operate in extremely demanding environments — high pressure, high temperature, abrasive slurries, corrosive chemicals, high rotational speeds, and critical safety applications. In such situations, engineering judgment built through years of experience remains irreplaceable.

This chapter examines both the genuine contributions that AI can make to seal engineering and the important limitations that engineers must understand.

13.2 The Role of Artificial Intelligence in Engineering

Artificial Intelligence essentially works by analyzing large volumes of data and identifying patterns. AI systems learn from historical performance data, failure databases, simulation results, and operational trends.

In engineering design, AI can support activities such as predictive maintenance, design optimization, performance forecasting, manufacturing process improvement, and condition monitoring. However, AI works effectively only when reliable and sufficient data exists.

In many mechanical seal applications — particularly in specialized industrial environments — sufficient structured data may not always be available. This is where engineering intelligence becomes far more valuable than any AI system.

13.3 Engineering Intelligence vs Artificial Intelligence

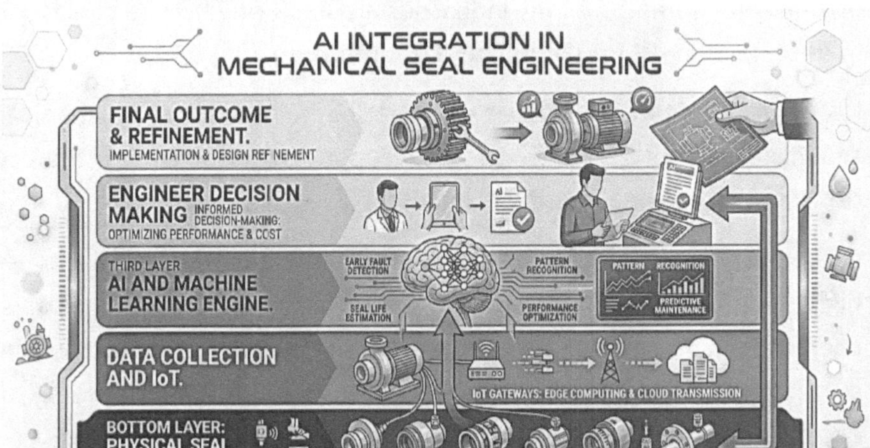

Understanding the difference between AI and engineering intelligence is essential for using both effectively.

Artificial Intelligence operates through pattern recognition, statistical learning, machine learning algorithms, and data correlations. It identifies relationships based on previously available information.

However, AI cannot fully understand physical intuition, failure mechanisms do not present in training data, unusual operating conditions, or creative engineering solutions. If a failure mode has never been seen before, no AI system can predict it.

Engineering Intelligence develops through deep understanding of physics, practical design experience, knowledge of materials and manufacturing, exposure to field failures, and analytical reasoning.

Engineering intelligence enables engineers to predict unknown problems, develop entirely new solutions, and innovate beyond existing data. It is the product of a career built on observation, analysis, and judgment.

AI supports Engineering Intelligence.
AI cannot replace Engineering Intelligence.
The most powerful engineering approach combines both.

13.4 AI Applications in Mechanical Seal Engineering

Artificial Intelligence can support mechanical seal engineering in several practical ways:

Failure Prediction

AI can analyse historical failure data to identify seal wear patterns, leakage trends, and operational anomalies. Machine learning models trained on large datasets of seal performance can provide early warning of developing failures, allowing maintenance to be planned before a breakdown occurs. This is particularly valuable in critical applications where unplanned downtime is costly.

Condition Monitoring

With the help of sensors and AI algorithms, seal performance can be monitored in real time. Parameters monitored may include temperature at or near the seal, vibration signature of the rotating equipment, pressure in the seal chamber, and leakage detection sensor outputs.

AI can identify early warning signals before a catastrophic failure occurs, moving maintenance strategies from reactive to predictive.

Design Optimization

AI tools can evaluate multiple design combinations quickly and suggest optimal seal face geometry, material combinations, and spring load configurations. However, the initial design principles must still come from engineers.

AI optimizes within a design space defined by human engineering knowledge; it cannot create that space.

Manufacturing Quality Control

AI-based inspection systems can improve manufacturing consistency by detecting surface defects, dimensional deviations, and assembly errors with speed and repeatability that exceeds human inspection. This helps improve reliability of seal components entering service.

13.5 Limitations of Artificial Intelligence in Seal Design

Despite its usefulness, AI has several important limitations when applied to mechanical seal engineering:

- **Lack of physical understanding** — AI systems do not inherently understand fluid dynamics, contact mechanics, thermal distortion, or tribological behaviour; they only interpret patterns in data
- **Limited experience in rare applications** — in many industrial sectors, failure data is limited or confidential, and without sufficient data, AI models cannot function effectively

- **Dependence on training data quality** — poor quality or biased training data produces unreliable AI predictions; the principle of garbage in, garbage out applies fully
- **Inability to innovate** — AI solutions depend heavily on historical data; engineering innovation often requires solutions that never existed before in any dataset
- **Black box problem** — many AI models, particularly deep learning systems, cannot explain why they make a particular recommendation, making engineering validation difficult

13.6 AI as an Engineering Assistant

The most productive way to use AI is to treat it as an engineering assistant rather than an engineering decision-maker. AI can assist engineers by processing large datasets, running simulations faster, identifying statistical trends, and automating repetitive calculations.

This allows engineers to focus on conceptual design, innovation, system-level thinking, and failure prevention. The ideal engineering approach is therefore Engineering Intelligence plus Artificial Intelligence — not one or the other.

13.7 AI and Reverse Engineering

Artificial Intelligence can also assist in the reverse engineering process. Possible AI applications include automated dimensional scanning and feature recognition, surface profile reconstruction and comparison, material property estimation from spectroscopic data, and digital twin modelling with AI-driven parameter identification.

However, AI cannot determine design intent — understanding the original design philosophy requires engineering interpretation, functional analysis, and knowledge of application conditions. This is the domain of the engineer, not the algorithm.

13.8 The Future of Engineering in the AI Era

The future of engineering will involve a strong collaboration between human engineering intelligence and artificial intelligence tools. Engineers who understand both engineering fundamentals and AI capabilities will be best positioned to develop innovative solutions. However, the foundation will always remain engineering knowledge.

Without deep understanding of mechanics, materials, thermodynamics, tribology, and fluid behavior, no intelligent design — artificial or human — can be achieved.

Young engineers entering the field today should embrace AI as a powerful tool, but should invest equally in building genuine engineering knowledge through experience, observation, and analysis.

The engineer who understands the physics deeply will always be able to guide, validate, and improve the work of any AI system. The engineer who relies on AI without physical understanding will be limited by the boundaries of the AI's training data.

13.9 Final Perspective

Artificial Intelligence represents an important technological advancement. But it must be clearly understood that AI can assist engineering — it cannot replace engineering intelligence. Particularly in complex engineering fields such as mechanical sealing, innovation arises from experience, technical insight, deep physical understanding, and creative thinking. These qualities define the true engineer.

> ***"AI can assist engineering —***
> ***it cannot replace Engineering Intelligence."***
> ***— Dr. Ashok Punde***

PART VI

Future of Mechanical Sealing Technology

Chapters 14 — 15

This page has been left blank intentionally.

Chapter 14

Future Trends in Mechanical Sealing Technology

14.1 Introduction

The mechanical sealing industry, like all engineering sectors, is subject to continuous change driven by evolving industrial requirements, new materials, digital technologies, and increasing environmental and sustainability pressures.

Understanding where the technology is heading allows engineers to prepare for the demands of tomorrow and to identify where engineering innovation is most needed.

This chapter examines the key trends shaping the future of mechanical sealing technology, with particular attention to those areas where the engineering principles discussed throughout this book will be most relevant.

14.2 Smart Sealing Systems

The concept of a smart seal — a mechanical seal that monitors its own condition and communicates performance data to an operator or control system — is moving from concept to industrial reality.

Smart sealing systems integrate miniaturized sensors directly into the seal assembly or seal chamber to measure temperature, pressure, vibration, leakage rate, and wear. The data from these sensors is transmitted wirelessly or through the equipment's instrumentation network to a monitoring system.

The engineering implications are significant. For smart seals to function reliably, the sensor integration must not compromise the mechanical integrity of the seal. The materials used for sensor housing must be compatible with process fluids.

The electronic components must withstand the operating temperatures and pressures of the application. These engineering challenges represent opportunities for innovation using the Ready Engineering and Re-Engineering methodologies described in earlier chapters.

14.3 Sensor-Integrated Seals

Sensor integration into mechanical seals requires careful engineering design to avoid creating new failure modes. Sensors must be positioned to measure meaningful parameters without disturbing the sealing interface.

Temperature sensors embedded near the seal face can provide real-time thermal data that indicates the lubrication condition. Acoustic emission sensors can detect the onset of dry running or face damage. Pressure sensors can identify cavitation or seal chamber pressure anomalies.

The engineering of sensor-integrated seals represents a true application of Ready Engineering — combining well-established seal design principles with emerging sensor technologies, adapted intelligently to the specific constraints of the sealing environment.

14.4 Digital Monitoring of Seal Performance

Beyond the seal itself, the broader digital monitoring of pump and rotating equipment performance is becoming standard practice in advanced industrial facilities. Continuous vibration monitoring, temperature trending, and flow measurement data can all be indicators of seal condition when properly interpreted.

The engineer's role in this environment shifts toward data interpretation and system-level diagnosis. The fundamental engineering knowledge described in this book — tribology,

materials science, thermal behavior, failure mechanisms — is what enables an engineer to extract meaningful information from monitoring data rather than simply observing numbers on a screen.

14.5 AI-Assisted Seal Diagnostics

As discussed in Chapter 13, AI-assisted diagnostics is one of the most actively developing areas in sealing technology. The combination of continuous sensor data, historical failure records, and machine learning algorithms creates the potential for highly effective predictive maintenance systems.

The most promising implementations will combine AI pattern recognition with engineering models that incorporate tribological and thermal physics, rather than relying on statistical patterns alone. Engineers with a deep understanding of seal behaviour will be needed to develop, validate, and refine these hybrid diagnostic systems.

14.6 Sustainability in Sealing Technology

Environmental sustainability is increasingly influencing mechanical seal design and selection. The drive to reduce fugitive emissions — leakage of volatile organic compounds and hazardous substances from rotating equipment — is one of the strongest forces shaping future seal design.

Regulatory requirements such as the API 682 standard and European emissions standards are setting increasingly stringent leakage targets.

From a materials perspective, the development of seal face materials that provide long life with minimal friction reduces both maintenance waste and energy consumption. From a system perspective, the trend toward zero-leakage double seal arrangements with environmentally safe barrier fluids reflects the industry's response to sustainability requirements.

Engineers who understand the complete seal system — not just the seal itself — will be best positioned to develop solutions that meet both reliability and sustainability objectives.

14.7 Customer-Friendly Seal Design Philosophy

One of the most important and often overlooked aspects of sealing technology development is the question of customer-friendliness. Too often in the industry, seal design has evolved in ways that increase the end-user's dependence on specialist suppliers, rather than empowering the user's own engineering capability.

A customer-friendly seal design is one that can be installed, maintained, and if necessary repaired by the end-user's own engineering staff without specialized proprietary tools or exclusive supplier support. It is one that uses materials that are available from multiple sources.

It is one whose performance is not critically dependent on the continuous involvement of the original supplier.

This philosophy is particularly relevant to the discussion of Dry Gas Seals for compressors — a technology that, despite its technical merits, has in many cases created over-dependence on a small number of specialized suppliers.

The pricing, lead times, and installation complexity associated with some proprietary seal systems create unnecessary burdens for end-users, particularly in developing industrial economies.

The engineering philosophy promoted in this book — Ready Engineering, Re-Engineering, and Reverse Engineering as tools for building genuine engineering capability — is fundamentally aligned with the customer-friendly design philosophy.

When engineers understand how seals work, they can make better decisions about selection, installation, and maintenance, and they are less dependent on supplier-driven recommendations.

The future of mechanical sealing technology should move toward designs that are more transparent in their operating principles, more accessible in their maintenance requirements, and more empowering to the engineers who work with them every day.

14.8 Integration of AI with Engineering Design

Looking beyond the near-term applications of AI in monitoring and diagnostics, the longer-term integration of AI with the engineering design process itself holds significant promise.

Generative design tools powered by AI can explore design spaces far larger than a human engineer could evaluate manually, potentially identifying non-obvious seal geometries or material combinations that offer performance advantages.

Simulation-driven design, where AI accelerates the evaluation of tribological and thermal models, can reduce the time from concept to validated design.

However, as emphasized throughout this book, these tools will only produce valuable results when guided and evaluated by engineers with genuine physical understanding. The engineer of the future must be both technically grounded in the fundamentals and comfortable with the digital tools that augment engineering practice. This combination — classical engineering

intelligence enhanced by modern computational tools — represents the most powerful engineering capability.

14.9 Conclusion

The future of mechanical sealing technology is rich with opportunity. Smart seals, digital monitoring, AI diagnostics, sustainable design, and customer-friendly engineering philosophy all represent directions in which the industry can and should develop.

The underlying engineering knowledge required to navigate these developments effectively is exactly what this book has sought to provide — a strong grounding in fundamentals, a clear understanding of the methodologies of Ready Engineering, Re-Engineering, and Reverse Engineering, and an appreciation of both the potential and the limitations of modern technological tools.

The next chapter addresses the ethical framework within which all engineering practice, including reverse engineering, should be conducted.

Chapter 15

Engineering Ethics and Intellectual Responsibility

15.1 Introduction

Engineering is a profession that carries significant responsibilities. Engineers make decisions that affect the safety of people, the protection of the environment, the efficiency of industry, and the advancement of technology.

This responsibility demands that engineering practice be guided not only by technical competence but also by ethical integrity.

In the context of this book — which addresses Ready Engineering, Re-Engineering, and particularly Reverse Engineering — the ethical dimension is especially important. Reverse engineering, in particular, operates in a space where the boundaries between legitimate engineering analysis and inappropriate appropriation of intellectual property must be clearly understood and respected.

15.2 Respect for Intellectual Property

Intellectual property (IP) in engineering takes many forms — patents that protect novel inventions, trade secrets that protect proprietary manufacturing processes, copyrights that protect technical documentation, and trademarks that protect product identities. All of these

represent investments of time, knowledge, and resources by the organizations and individuals who created them.

Engineers who work with existing products — whether in re-engineering or reverse engineering activities — must understand and respect these protections. Studying a product to understand how it works does not automatically create the right to reproduce it commercially.

If a seal design is protected by a valid patent, manufacturing and selling that design without a license is an infringement, regardless of how the dimensions and materials were determined.

Responsible engineers seek legal advice when uncertain about IP boundaries, and they design their activities to respect the intellectual effort of others even when not strictly required by law to do so.

15.3 Patents and Their Role in Engineering Innovation

The patent system exists to encourage innovation by providing inventors with a time-limited exclusive right to their inventions in exchange for publicly disclosing how those inventions work. The disclosure requirement is important: patent documents are a rich source of engineering knowledge that is entirely legitimate and intended for public use.

Reading and studying patent documents is an important part of engineering practice — it is how engineers learn from the published work of others, understand the state of the art, and identify opportunities for further innovation.

What is not permitted is manufacturing and selling a patented product during the term of the patent without a license. Engineers must be aware of relevant patents in their field and must ensure that their designs do not infringe valid patent claims.

This requires an understanding of both the technical content of patents and the legal framework within which they operate.

15.4 Ethical Reverse Engineering — Principles and Boundaries

Reverse engineering, as practiced legitimately and ethically, involves the systematic analysis of an existing product to understand its design principles. The purpose may be to develop compatible replacement parts, to understand the state of the art in a technical field, to investigate a failure, or to develop an improved alternative design.

All of these are legitimate engineering activities.

The ethical boundaries of reverse engineering can be understood through several principles:

- **Purpose** — is the reverse engineering activity aimed at genuine technical understanding and improvement, or primarily at reproducing a proprietary product to compete commercially without investing in original development?
- **Improvement** — does the reverse engineering activity lead to a design that represents genuine engineering advancement over the original, or merely a functional copy at a lower price?
- **Transparency** — is the engineer prepared to explain and defend the reverse engineering methodology used, or would that explanation be professionally embarrassing?
- **Safety** — does the resulting design meet the safety and performance requirements of the application, and has it been properly validated?

An engineer who can answer all four of these questions honestly and positively is working within the ethical framework of the profession.

15.5 Innovation vs Imitation — The Critical Distinction

The distinction between innovation and imitation is the central ethical question in reverse engineering. It is also the central message of this book.

Imitation seeks to reproduce what already exists as closely as possible, contributing nothing new to the engineering knowledge of the field. It exploits the investment of others without adding value. Over time, an industry that consists primarily of imitators stagnates, because no one is investing in genuine advancement.

Innovation seeks to understand what exists and then move beyond it — to create something better, more reliable, more economical, or more suited to the needs of users. Innovation requires the deeper engagement with engineering principles that this book describes.

It requires the engineer to ask not just what the original design was, but why, and whether there is a better answer.

The engineer who approaches reverse engineering with the intention of genuine innovation will invariably produce better outcomes — better products, better engineering knowledge, and better professional reputation — than the one who is satisfied with imitation.

> ***Imitation reproduces the past.***
> ***Innovation creates the future.***
> ***Engineering Intelligence is the difference.***

15.6 Engineering Responsibility in the Age of AI

The emergence of AI tools in engineering creates new ethical dimensions. AI systems can generate designs, predict failures, and optimize parameters at speeds and scales that no human engineer can match. But they do so without moral agency, without professional accountability, and without the contextual understanding that human engineers possess.

This means that the engineer who uses AI tools carries full professional responsibility for the outputs those tools produce. An AI system cannot be held professionally responsible for a failed seal design. The engineer who selected, configured, applied, and validated the AI's recommendations is responsible. This responsibility cannot be delegated to a software tool.

Engineers in the AI era must maintain the technical competence to evaluate and validate AI outputs. They must not use AI tools as a substitute for their own engineering judgment but as an extension of it. And they must be prepared to override AI recommendations when their engineering knowledge indicates that the recommendation is inadequate or inappropriate for the specific application.

15.7 The Responsibility of Engineers Toward Young Professionals

Experienced engineers have a particular responsibility toward the next generation of professionals entering the field. This responsibility has several dimensions.

The first is mentorship — sharing the practical knowledge, field experience, and engineering judgment that cannot be learned from textbooks alone. Much of what makes an experienced seal engineer effective is tacit knowledge: the ability to look at a failed seal face and immediately recognize the failure mode, or to listen to a pump and sense that something is wrong with the seal.

This knowledge must be passed on through direct engagement.

The second is honesty about the complexity of the field. Young engineers are sometimes discouraged from reverse engineering or from questioning established designs because these activities are associated with copying.

The engineering profession has a responsibility to communicate clearly that reverse engineering, conducted with proper knowledge and ethical integrity, is a legitimate and valuable engineering discipline.

The third is encouraging genuine engineering curiosity. The engineer who asks why a design was made this way, who investigates failures with genuine interest rather than frustration, who reads patent documents and technical papers outside of immediate project requirements — this engineer will develop the engineering intelligence that drives true innovation.

15.8 Conclusion

Engineering ethics is not a constraint on engineering practice — it is the foundation of it. The engineer who works with integrity, who respects the intellectual effort of others, who uses reverse engineering as a tool for genuine learning and improvement rather than mere reproduction, and who accepts full responsibility for the safety and performance of the designs they produce, is practicing engineering at its highest level.

The technical methodologies described throughout this book — Ready Engineering, Re-Engineering, and Reverse Engineering — are most powerful when applied within this ethical framework. Engineering intelligence, combined with ethical integrity, is the most valuable professional asset any engineer can possess.

Conclusion

Engineering Intelligence in the Age of Artificial Intelligence

This book has explored one of the most important and yet frequently misunderstood areas of engineering practice — the application of Ready Engineering, Re-Engineering, and Reverse Engineering to mechanical sealing technology, in an era where Artificial Intelligence is reshaping the tools available to engineers.

The journey through the fifteen chapters of this book has covered the fundamentals of engineering and sealing technology, the principles and practice of mechanical seal design, the methodologies of re-engineering and reverse engineering, the tribological and material science foundations of seal performance, the tools of modern engineering practice, and the ethical framework within which engineering must be conducted.

At every stage, one theme has recurred: Engineering Intelligence — the accumulated understanding of physics, materials, operating conditions, failure mechanisms, and design principles that is built through years of disciplined practice — is irreplaceable. It cannot be generated by a data-driven algorithm.

It cannot be downloaded from a software system. It is built by engineers who observe carefully, analyse rigorously, and never stop asking why.

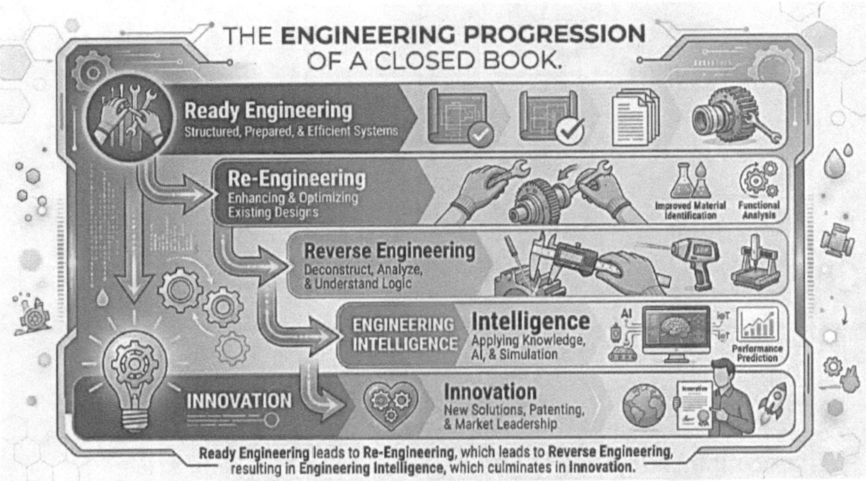

The Engineering Knowledge Ladder that appears at the beginning of this book and recurs throughout it represents a genuine progression. Every engineer who studies this book and applies its principles will move along this ladder.

The goal is not merely to be competent at each stage, but to develop the integrated understanding that makes true engineering innovation possible.

Reverse Engineering, when approached with engineering intelligence and ethical integrity, is not imitation. It is the most demanding form of engineering analysis — the requirement to understand an existing design so thoroughly that you could not only reproduce it, but improve it.

The engineer who can do this has achieved a level of understanding that transcends the original design.

Artificial Intelligence will become an increasingly important part of the engineering toolkit. The engineers who use it most effectively will be those who understand the physics deeply enough to know when to trust the AI's output and when to override it. Engineering intelligence guides the use of AI, not the reverse.

> ***"Reverse Engineering without Engineering Intelligence becomes Copying.***
> ***Reverse Engineering with Engineering Intelligence becomes Innovation."***
>
> ***— Dr. Ashok Punde***

To the engineers who read this book — whether at the beginning of their careers or in the later years of a long professional journey — the message is this: invest in your engineering knowledge.

Understand the physics of what you design. Study failures with the same intensity that you bring to successful designs. Approach every existing product as an opportunity to learn. And always, always ask why.

The mechanical sealing industry needs engineers of this kind — engineers who combine deep technical knowledge with the intellectual courage to innovate, and the ethical integrity to do so responsibly.

Engineering progress has never come from copying what already exists. It comes from understanding, improving, and creating something better.

Dr. Ashok Punde
Independent Consultant — Static & Dynamic Sealing Technology

This page has been left blank intentionally.

Glossary of Technical Terms

The following terms are used throughout this book. Definitions are provided in the context of mechanical sealing engineering.

Abrasive Wear

Wear caused by hard particles or surface asperities that cut or scratch the seal face material during sliding contact.

Balanced Seal

A mechanical seal design in which the hydraulic closing force on the seal faces is reduced by geometric design, allowing operation at higher pressures with lower face loading and heat generation.

Boundary Lubrication

A lubrication regime in which the fluid film between seal faces is very thin and significant direct surface contact occurs. Typical during start-up and low-speed operation.

Carbon Graphite

A commonly used seal face material with excellent self-lubricating properties, used as the softer face in many seal combinations.

Cartridge Seal

A pre-assembled mechanical seal unit in which all components are factory-set and pre-aligned, simplifying installation and reducing the risk of installation error.

Contact Pressure

The average pressure exerted between the seal faces, determined by the spring closing force and hydraulic forces acting on the seal cross-section.

Digital Twin

A virtual model of a physical component or system, updated with real operating data, used for performance monitoring and predictive maintenance.

Double Mechanical Seal

A seal arrangement using two mechanical seals installed in series, with a barrier fluid circulating between them. Used for hazardous or toxic process fluids.

Dry Running

Operation of a mechanical seal without adequate fluid film lubrication between the seal faces, causing rapid heat generation and face damage.

Dynamic Seal

A sealing device designed to prevent leakage where relative motion exists between the sealed components, such as a mechanical seal on a rotating shaft.

Elastomer

A rubber-like polymer material used for secondary sealing elements such as O-rings. Common types include NBR, FKM, EPDM, and PTFE.

Engineering Intelligence

The accumulated understanding of physics, materials, failure mechanisms, and design principles developed through years of practical engineering experience.

Failure Pyramid

A conceptual model illustrating the relative contribution of different failure causes (installation, alignment, application, materials, design) to mechanical seal failures.

Flush Plan

A standardised piping arrangement (per API 682) used to provide lubrication, cooling, or environmental control to a mechanical seal chamber.

Flatness

A measure of how closely a seal face approximates a perfect plane, typically expressed in helium light bands (1 light band ≈ 0.3 micrometres). Critical for effective sealing.

Hydrodynamic Lubrication

A lubrication regime in which a continuous pressurised fluid film completely separates the seal faces, minimising friction and wear.

Mechanical Seal

A device used to prevent leakage of process fluid along a rotating shaft, using controlled contact between two precisely flat sealing faces.

Mixed Lubrication

A lubrication regime in which both direct surface contact and fluid film lubrication occur simultaneously at different locations on the seal face.

Ready Engineering

An engineering methodology in which new designs are developed by combining fundamental engineering principles with proven design knowledge and established industrial practices.

Re-Engineering

The systematic process of analysing an existing engineering design and introducing modifications to improve its performance, reliability, or efficiency.

Reverse Engineering

The systematic process of examining and analysing an existing product to understand its design structure, materials, dimensions, and functional principles, with the aim of learning, improving, or adapting the design.

Seal Balance Ratio

The ratio of the hydraulic closing area to the opening area of a mechanical seal face, used to control the hydraulic face pressure.

Seal Face

One of the two precision-lapped flat surfaces that form the primary sealing interface of a mechanical seal.

Silicon Carbide

A hard ceramic material widely used for mechanical seal faces due to its high hardness, thermal conductivity, and chemical resistance.

Static Seal

A sealing device used between two stationary components, such as a gasket or O-ring in a flanged connection.

Tribology

The science and engineering of interacting surfaces in relative motion, encompassing friction, wear, and lubrication.

Tungsten Carbide

A very hard composite material used for mechanical seal faces in abrasive applications.

Acknowledgments

A Reflection of Gratitude Beyond Words

This Book, ***"Mechanical Seals – Ready Engineering, Re-Engineering, and Reverse Engineering"*** is not merely a compilation of knowledge—it is a journey lived, a path walked, and a silent dialogue between experience and understanding.

Every page carries not only technical insight but also the unseen imprints of time, learning, challenges, and realization. What appears as engineering on the surface is, in truth, a deeper expression of observation, patience, and inner evolution.

At the Heart of This Journey Stands My Ever-Guiding Mentor, The Late Mr. Ajitsingh Bhamvra. *His wisdom was not confined to words, it was reflected in the way he saw problems, approached solutions, and nurtured clarity in complexity. His guidance continues to echo within my thoughts, like a steady light that neither fades nor falters. To have learned under him was not just an opportunity, it was a defining blessing.*

My Gratitude also flows to the Countless Colleagues, Professionals, and Experts from the Mechanical Seal industry, with whom I shared nearly five decades of professional life. Each interaction, each challenge, and each solution contributed quietly yet powerfully to this collective expression. ***What I have written here is, in many ways, a reflection of what we have all lived together.***

I extend my sincere appreciation to **Mr. Madhu Nair And His Publishing Team**, whose support transformed an inner intent into a tangible creation. Their trust and encouragement allowed this work to take form, much like a design finding its purpose through thoughtful execution.

To my Family and Friends—your presence has been the silent strength behind this work. Your patience during long hours, your belief during moments of doubt, and your unwavering support have been the foundation upon which this journey stood firm. You have been the balance in my motion, the calm within my pursuit.

I also bow in respect to all those known and unknown minds whose innovations, thoughts, and explorations in design thinking and engineering have shaped the path for many of us. We are, after all, continuations of a larger flow of knowledge.

As I place this work before the reader, I do so with humility and hope—that it may not only serve as a technical guide but also inspire a deeper way of thinking, observing, and creating. Because true engineering is not just about making things work—it is about understanding why they exist, and how they can evolve.

Dr. Ashok Punde
Pune, India

About the Author

Dr. Ashok Punde is a distinguished professional in the field of Mechanical Sealing Technology, with over 48 years of experience spanning design engineering, product development, industrial leadership, and technical consulting.

He began his career as a Mechanical Seal Designer, gaining deep hands-on experience in the design and development of sealing systems for a wide range of industrial applications. Over the years, he expanded his expertise into manufacturing engineering, materials technology, product development, and strategic business management.

During his distinguished career, he served as Managing Director and Chief Executive Officer for over 23 years, where he led organizations involved in mechanical sealing solutions, driving innovation, product development, and market expansion.

During this period he worked across a wide spectrum of sealing applications, including pumps, compressors, agitators and mixers in process industries including chemical processing, petrochemicals, power generation, pharmaceuticals, and water treatment, as well as highly demanding applications in aviation ground equipment, slurry pumping for metal-producing refineries, and specialized engineering sectors involving critical reliability requirements.

Currently, he operates as an Independent Consultant, providing expertise in Static and Dynamic Sealing Technology, Mechanical Seal Product Development, Reverse Engineering and Re-Engineering, Industrial Problem Solving, and Business Enhancement and Strategic Mentorship.

His work focuses on helping companies improve product performance, reliability, and engineering capability, while mentoring the next generation of engineers in the practical and intellectual foundations of sealing technology.

Through this book, Dr. Ashok Punde shares the practical engineering philosophy and industry knowledge gained over nearly five decades in the mechanical sealing field, with the particular aim of establishing that Reverse Engineering, when practiced with genuine engineering intelligence and professional integrity, is a discipline of innovation — not imitation.

Written By *Dr. Ashok Punde*
Consulting Professional & Strategist In Static & Dynamic Sealing | Mentor & Purposeful Communication Architect
E-Mail : acp@susasconsultech.in

www.ingramcontent.com/pod-product-compliance
Lightning Source LLC
LaVergne TN
LVHW100316210926
848480LV00003B/1142
9798905408403